THE TRIUMPH OF THE EMBRYO

Lewis Wolpert

With illustrations drawn by

Debra Skinner

OXFORD   NEW YORK   TOKYO
OXFORD UNIVERSITY PRESS

Oxford University Press, Walton Street, Oxford OX2 6DP

Oxford New York Toronto
Delhi Bombay Calcutta Madras Karachi
Kuala Lumpur Singapore Hong Kong Tokyo
Nairobi Dar es Salaam Cape Town
Melbourne Auckland Madrid

and associated companies in
Berlin Ibadan

Oxford is a trade mark of Oxford University Press

Published in the United States
by Oxford University Press Inc., New York

© Lewis Wolpert 1991
First published 1991
Reprinted (with corrections) 1992

First published in paperback 1993

A catalogue record for this book is available
from the British Library

Library of Congress Cataloging-in-Publication Data
Wolpert, L. (Lewis)
The triumph of the embryo/Lewis Wolpert; with illustrations drawn by Debra Skinner.
1. Embryology, Human—Popular works. I. Title.
QM603.W65   1991   612.6´4—dc20   91–7583
ISBN 0–19–854243–7 (hbk)
ISBN 0–19–854799–4 (pbk)

Printed in Great Britain by
Butler and Tanner Ltd, Frome, Somerset

THE TRIUMPH OF THE EMBRYO

Lewis Wolpert, CBE, FRS, is Professor of Biology as Applied to Medicine at University College London. As a broadcaster and distinguished speaker, he is committed to communicating the excitement of doing and thinking about science to the widest possible audience. This was the theme of his previous book for Oxford University Press, *A passion for science*.

PREFACE

I HAVE WANTED to write this book for a long time. There is no other book which enables non-specialists to learn about how embryos develop. This is very unfortunate, for the process of embryonic development is one of the most exciting problems in modern biology. One could say that together with trying to understand how the brain works, they are the great biological problems of our time. The problem of development is how a single cell, the fertilized egg, gives rise to all animals, including humans. So it really is about life itself. Even those of us who work on these problems seldom lose a sense of wonder at this remarkable process.

The origin of this book lies in the Christmas Lectures I gave at the Royal Institution in 1986 under the title 'Frankenstein's quest'. However, the material has been very substantially altered. It must be recognized that while I have tried to give a picture of the field as it is, it is, nevertheless, a personal view. My selection of examples, and the emphasis I have given to different processes, will no doubt be different from other workers in the field. Perhaps the strongest prejudice I have is that there are unifying principles to account for the way embryos develop; that at some deep level there are only a few basic mechanisms that are used in the development of all animals. It is too soon to know if I am correct. How animals develop still presents many problems. I do not want to give the impression that we understand more than we in fact do—it is a very active and rapidly moving field of research.

Most of the book is devoted to embryonic development, but there is more to what is now called developmental biology than embryonic development. Essentially the same mechanisms that are used in embryos are also used by those animals that can, for example, regenerate their limbs. Thus there is a chapter on regeneration. Similarly, there are chapters on growth and ageing, since both are related to embryonic development. Again, there is a short chapter on cancer, treating it as an abnormal developmental process. Finally, the role of development in evolution is given some attention.

How difficult is the book for the layman or non-biologist? Quite easy I hope. There is inevitably a new vocabulary to be learned—the reader will have to come to terms with gastrulation, neurulation, differentiation, induction, and others—but not too many. There are new ideas that may be a little troublesome at first. But there are really rather few key ideas that have to be mastered. No more difficult are those parts of the book that deal with cells at the molecular level. Some appreciation of DNA and proteins is necessary and I have provided the necessary background where appropriate. Probably the greatest difficulty, and one I have not overcome, is in trying to visualize the changing shape of the developing embryo. This is hard enough for professionals and is often like trying to describe how to tie a shoelace. While direct observation of the embryo is the ideal, I hope the drawings and photographs will be a great help.

For those who wish to read further in the subject, I strongly recommend the chapters on development in *The molecular biology of the cell* by Alberts *et al.* (Garland 1989). Another good textbook is *Developmental biology* by Scott Gilbert (Sinauer 1988).

The following people have provided the photographs which appear in this book; for their generosity and cooperation I thank Amata Hornbruch, Kathleen Sulik, Dennis Bray, Julian Lewis, Charles Brooke, and Peter Lawrence. I am indebted to many people. Maureen Maloney, my secretary, typed innumerable drafts, made countless corrections, and provided essential encouragement. Mary and Jack Herberg, William Graves, Patti Suzman, and Cheryll Tickle read drafts and made valuable comments. Judy Hicklin and Alison Richards pointed out shortcomings and provided invaluable suggestions for improvement. At Oxford University Press my editor was patient, firm, and always helpful. To all I am very grateful.

University College London   Lewis Wolpert
December 1990

CONTENTS

I. CELLS AND EMBRYOS   1
II. MOULDING OF FORM   11
III. PATTERN FORMATION   31
IV. FINGERS AND TOES   59
V. EX DNA OMNIA   77
VI. CELL DIVERSITY AND DIFFERENTIATION   91
VII. GENES AND FLIES   105
VIII. WIRING THE BRAIN   119
IX. SEX   135
X. GROWING   145
XI. CELL MULTIPLICATION AND CANCER   157
XII. AGEING   165
XIII. REGENERATION   171
XIV. EVOLUTION   183
XV. A PROGRAMME FOR DEVELOPMENT   199
INDEX   205

CELLS AND EMBRYOS

Wʜᴀᴛ ᴄᴏᴜʟᴅ be more important or intriguing than our own origins? Like all animals we come from one cell that develops into an embryo which forms the adult. This embryonic development presents a fundamental problem of biological organization. From the single cell, the fertilized egg, come large numbers of cells—many millions in humans—that consistently give rise to the structures of the body. How do these multitudes of cells become organized into the structures of, for example, our body—nose, eyes, limbs, and brain? What controls their individual behaviour so that a global pattern emerges? And how are the organizing principles, as it were, embedded or encoded within the egg? It is remarkable that a cell as overtly dull and structureless as the fertilized egg can give rise to such varied and complex forms. The answer lies in cell behaviour and how this behaviour is controlled by genes.

Genes control development. Genes are specific parts of the DNA in the chromosomes that all animals inherit from their parents. Genetics deals with how genes are passed on from parents to their offspring and a great deal is known about the mechanisms governing this process. Very much less well understood is how the genes control embryonic development. To link genes with fingers and brains is the

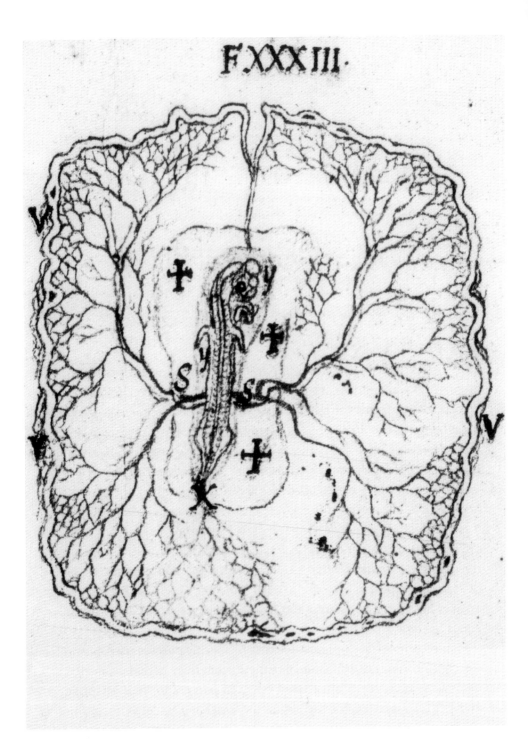

Drawing of three-day chick embryo by Marcello Malpighi in 1672

enormous challenge that the problem of development presents. To understand how the development of different structures is controlled by genes, how the genes in the fertilized egg of a sea-urchin, frog or chimpanzee result in the emergence of that animal.

We know that changes in genes—mutations—can alter the colour of our eyes, can cause the development of extra fingers, and can result in flies with a leg growing out of the head instead of an antenna. What is not known is how the genes exert these effects during development. We want to know the relationship between genes and structures in the body; are there, for example, special genes for the arm? Is there a gene for each nerve cell? To put the problem in an extreme form, if we had access to all the genetic information in the egg and knew all the genes in detail, could we compute the animal to which it will give rise?

ORIGINS

Ideas, in a scientific sense, about the nature of development, started with Hippocrates in the fifth century BC. His understanding was couched in terms of fire and humidity, wetness and solidification. But at least it was a start towards trying to find a mechanism. Embryology proper really began a century later with Aristotle defining a basic question which was to dominate the field until the end of the nineteenth century. Do, Aristotle asked, all the parts of the embryo come into existence together, or do they appear in succession? Is everything preformed from the beginning, or is development more like the knitting of a fisherman's net? He thus defined the preformation/epigenesis controversy.

Aristotle favoured the knitting metaphor, which he termed epigenesis. His rejection of preformation was on philosophical and experimental grounds. Since he believed the embryo was formed from the menstrual blood by the male dynamic in the semen it was clear to him that preformation must be excluded. He also opened fertile chicken eggs and concluded, falsely, that the heart was the first organ to develop. In the end he was correct about epigenesis, but for quite the wrong reasons.

Aristotle's influence was enormous, and even the great English physician, William Harvey, who laboured so long on the chick embryo, could not shake it off. Like Aristotle, he supported epigenesis but with as little reason.

The contrary view, that all embryos existed from the beginning of the world, was first formulated in the 1690s by, among others, Malebranche in France and Swammerdam in the Netherlands. These embryologists could not accept the explanation of epigenesis provided by the French philosopher Descartes, who suggested that there were mechanical forces which moulded the embryo. It seemed to them naïve to suppose, like Descartes, that physical forces could lead to the formation during embryonic development of such complex structures as animals possessed, particularly since Descartes in no way explained how the forces could mould the embryo. They believed, by contrast, that all the parts of the embryo were preformed from the very beginning and merely grew during embryonic development and so gradually became visible. In their view, the first embryo of a species contained all future embryos. To Malebranche, for example, it appeared not unreasonable that there were an infinite number of trees in a single seed; this, he argued, would only seem extravagant to those who measured God's powers by their own imagination. Even Malpighi, the brilliant Italian biologist, could not break free of such ideas. While he provided a remarkably accurate description of the development of the chick, he remained convinced, against his own evidence, that the embryo was present before the egg was laid. Because the parts, early on, are so small, he thought that they could not be observed, even with his most powerful microscope.

The eighteenth-century preformationists had an answer to all criticisms. When the Frenchman, Charles Bonnet, was confronted with the argument that if preformation were true the first rabbit would have had to contain $10^{10\,000}$ preformed embryos, he merely responded by saying that it was always possible, by adding zeros, to crush the imagination under the weight of numbers. For him preformation was 'one of the greatest triumphs of rational over sensual conviction'. For their part, the problem faced by the proponents of epigenesis was that no one could provide a reasonable explanation of how the embryo was formed. Proponents of epigenesis could only invoke a 'master-builder' within the embryo, or a *vis vitalis*.

Experiments and observations failed to resolve the controversy. And that was for a very good reason. It was, in principle, insoluble without an understanding of cells and genes. Only in the latter half of the last century did these ideas begin to emerge and it was recognized that embryos developed by epigenesis. As we shall see, DNA provides the programme which controls the development of the embryo and brings about epigenesis.

CELLS

Cells are the basic units of life. They are the true miracle of evolution. Miracle in the figurative sense, since although we do not know how cells evolved, quite plausible scenarios have been proposed. Miraculous, none the less, in the sense that they are so remarkable. Most remarkable, and, in a way, a definition of life, is their ability to reproduce themselves. They are able to take in chemicals and convert them into usable energy and to synthesize all the components of the cell during growth that eventually leads to cell multiplication.

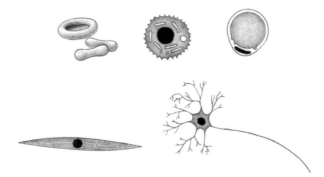

Animals are made up of specialized cells, such as blood cells, cartilage cells, fat cells, muscle cells, nerve cells—humans have about 350 different cell types while lower animals, like hydra, only 10 to 20. Cells carry out an amazing range of specialized functions, such as carrying

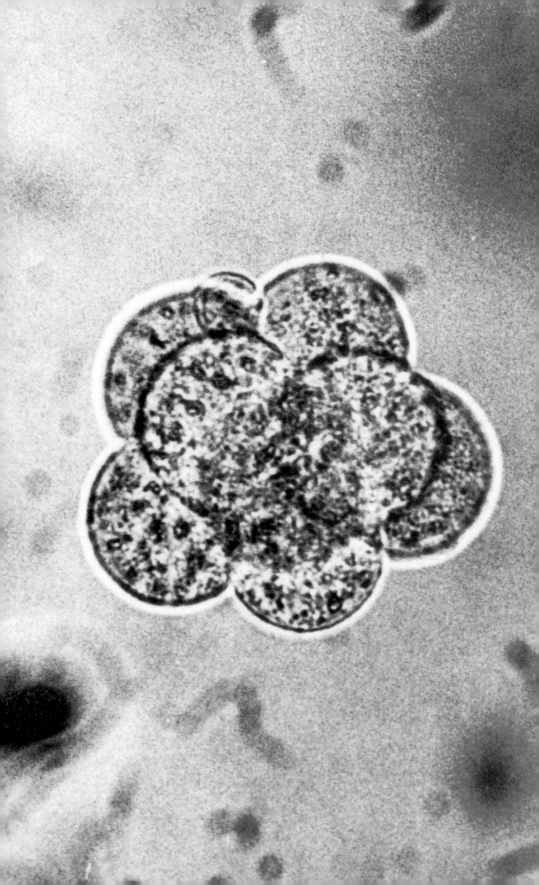

oxygen, transmitting messages, contracting, secreting chemicals, synthesizing molecules, and multiplying. The cells in the embryo are initially much less specialized and differ from each other in more subtle ways. All have certain basic characters and in order to understand their role in development it is helpful to be aware of four cell activities, three cell structures, and two main kinds of molecule.

The four cell activities are cell multiplication, cell movement, change in character, and cell signalling. These are mainly self-explanatory. Cells multiply by dividing and this usually requires cell growth, the cells doubling in size before dividing in two. Cells can also change shape, exert forces, and move from one place in the embryo to another. They can also change character: during development cells change from having rather unspecialized characters to mature cells with very specific functions. Cells in different parts of the embryo can be thought of as developing along quite different pathways, diverging more and more in character. Finally, cells give out and receive signals from neighbouring cells.

Cells cannot normally be seen without a microscope, being about one-thousandth of a millimetre in diameter. However, some cells, like the large eggs of frogs, are easily visible, and the human egg is just visible to the naked eye. The three key cell structures are the cell membrane, the cytoplasm, and the nucleus. Surrounding the cell is a very thin cell membrane which controls the entry and exit of molecules, and maintains the integrity of the cell. On the surface of the membrane are special receptors for signals from other cells, as well as molecules that enable cells to adhere to one another. Confined by the cell membrane the cytoplasm contains all the machinery for production of energy and cell growth; and there are also structures in the cytoplasm which can generate forces to change the shape of the cell, resulting in cell movement. Embedded within the cytoplasm is the cell nucleus surrounded by its own special membrane. Within the nucleus are the chromosomes of the cell, which contain the genes.

The life of the cell is dependent on the chemical reactions among the many million constituent molecules. Two key classes of molecules are nucleic acids and proteins which will be described much more fully in

An eight-cell stage mouse embryo

Chapter 5 and can be largely ignored for the present. Genes are made of the nucleic acid DNA, and they exert their effect by determining which proteins are made in the cell. Proteins are fundamental to the life of the cell because they are essential for all the key chemical reactions as well as providing the main framework of all the structures in the cell. Almost all the chemical reactions in the cell such as the provision of energy or the synthesis of key molecules will only take place in the presence of special proteins, known as enzymes, which allow the reactions to occur. Proteins are also the major structural molecules in the cell, providing, for example, the forces for cell movement, receptors at the cell surface, and the adhesive links between cells. Proteins also give different cells their special character. For example, the protein haemoglobin carries oxygen in the red blood cells, and insulin, another protein, is made in particular cells in the pancreas.

The wide variety of proteins in the cells are all coded for by the genes in the nucleus. While proteins themselves are synthesized in the cytoplasm, whether or not a protein is made is dependent on whether or not the gene that contains the information for that protein is active (Chapter 5). The only function of genes is to determine which proteins are made, thereby determining which chemical reactions take place and which structures will be present in the cell. In this way, genes control cell behaviour.

Cell behaviour can thus provide the crucial link between genes and animal structure and form. If we can understand how cells behave during development so as to make arms and brains we can then begin to ask how genes control the behaviour of the cells and so establish the link. Cells thus provide the key to understanding development because their behaviour brings about embryonic development and is controlled by gene activity. In very general terms there are, in development, three kinds of genes—those that control spatial organization, those that control change in form, and those that control cell differentiation.

THE MODERN ERA

We are now in the midst of a most exciting stage in cell and developmental biology. The revolution that resulted in molecular biology enabled us to begin to understand many of the processes in the cell at the molecular level, although it must be stressed that we have a long way to go. The fundamental achievement was the elucidation of the structure of DNA and proteins together with the mechanisms by which genes code for and control the structure of proteins. Part of the beauty of this revolution was the recognition that at the level of genes and proteins the mechanisms used by all animals are almost universal. Developmental biology also hopes to reveal such general mechanisms.

The powerful techniques of molecular biology are now being applied to the classical problems of embryonic development that were studied by an older generation of biologists. Their work laid the foundation for understanding developmental mechanisms and was largely descriptive, carefully recording the changes in the form of the embryo during both normal development and when it was experimentally perturbed. Many of the experimental perturbations involved rearranging or removing the parts of the early embryo and seeing what effect this had on the embryo's development. These experiments led to theories about how development was controlled in terms of cell and tissue properties, but it was very difficult to link these theories with gene action. The modern approach is descriptive too, but at the molecular level. More important, the modern approach can lead to explanations at the molecular level thus linking gene action to developmental process.

The emphasis of the molecular level of description requires some justification. Development of the embryo can be described, and explained, at various levels of organization such as that of the tissue, cell, or molecule. This is true of many biological processes and no one level is intrinsically preferable to another. One would not try to explain the flight of birds at the molecular level, whereas that level is most appropriate if we wish to understand how muscles contract. For development, the importance of explanations at the molecular level is that genes act at this level. That is, genes control the synthesis of protein molecules, and, for that reason, there is general acceptance that it is at the level of molecular interactions that the behaviour of cells and embryos needs to be understood. Molecules are the natural language of the cell. Another reason is that when processes are understood at the molecular level then there is at once a natural link to chemistry, a most powerful body of knowledge to advance further understanding. While explanations at the molecular level are the aim, much of this book will still look at embryos at the cellular level, reflecting the current state of knowledge.

In order to understand how embryos develop, it is necessary to do experiments on the embryo itself, which may involve opening chicken eggs and observing their development as Aristotle did, grafting parts from one region to another, or adding specific chemical agents to observe their effects. Embryologists are sensitive about issues relating to animal experimentation and the vast majority of experiments are done on very early embryos which are rarely allowed to develop to maturity. Of crucial importance is that none of these experiments require the animals to suffer any pain whatsoever. The vast majority are performed long before a nervous system is present and in the small number of cases where this is not so, the animal is anaesthetized.

Can we expect to find the fundamental and universal principles in development? Or is development just the accretion of a variety of different mechanisms? I find it hard to believe that so fundamental a process is not governed by principles just as elegant and universal as those uncovered in relation to molecular genetics. Once evolution had discovered successful ways of constructing organisms it would surely have used those same mechanisms again and again. This book is based on that conviction; it has the crucial implication that understanding how one kind of animal develops will help us understand the development of all the others.

I will show that there is no 'master builder' in the embryo, no vital force. Each cell has its own developmental programme which makes use of a limited number of processes that have been used again and again, for hundreds of millions of years.

MOULDING OF FORM

FORM EMERGES due to changes in cell shape and these provide direct evidence for epigenesis. Development begins with the fertilized egg, which is a single cell, giving rise to a number of smaller cells. Cell divisions cleave the egg, like cutting a cake, and result in a multi-cellular structure. These cell divisions simply divide up the egg to give a population of smaller cells that form the early embryo and are thus different from the cell division associated with cell growth and multiplication. Because growth is not required the early cleavage divisions typically transform the egg, within a few hours, into a hollow ball with the cells arranged as a spherical sheet. This simple structure, the blastula, must now be moulded by cellular activities into all the shapes that emerge during development.

Eggs come in a wide range of sizes depending on how much yolk they have. Chicken eggs are large because of the yolk which acts as a source of nutrients for the growth of the developing embryo. The chick embryo proper comes from a very small region resting on the yolk and which is equivalent to the mammalian egg. The human and mouse eggs are about one tenth of a millimetre in diameter and are just visible to the naked eye. Unlike the chick they contain no yolk and their growth is dependent on nutrients being supplied by the mother. Frog eggs are about one to two millimetres in diameter and, like the chick, have enough yolk to enable them to grow to a stage when they can feed themselves.

The blastula gives no visible indication of the complex animal into which it will develop. It is only after the next stage, gastrulation, that the form of the animal begins to emerge.

GASTRULATION

I have been quoted as saying that it is not birth, marriage, or death, but gastrulation which is truly 'the important event in your life'. Excessive perhaps, but the occasion was the attempt to convince the clinician of the importance of studying early development.

Gastrulation occurs in the development of all animals. It is the process that occurs when the cells of the blastula rearrange and move so that the hitherto rather simple, and often spherical or flat embryo is transformed into something approaching the form from which the animal will develop. During gastrulation the front and back, top and bottom become evident, and the basic body plan is laid down. Also, the cells move so as to take up new positions. The development of the gut illustrates the point. Surprisingly, cells that will form the gut are on the outside surface of many early embryos. However, the gut in all animals is an internal structure. It is during gastrulation that the cells leave the outer surface and move inside the region where they will form the gut. Similarly, in early vertebrate embryos, the cells that will form the vertebral column and the muscles are also on the outside and must move inwards to an appropriate location. It is only after gastrulation that the organs, like limbs, liver, and eyes, begin to develop.

In most vertebrates in which they have been intensively studied—amphibians and birds—gastrulation movements are rather complicated. Movements occur simultaneously over many parts of the embryo with sheets of cells streaming past each other, contracting and expanding. It taxes the minds of determined embryologists to try and visualize what is going on. Fortunately, there are animals in which the process is not only much simpler but can be observed directly.

We can follow gastrulation in the early development of the sea-urchin and actually see the gut forming. Sea-urchin embryos are very attractive for developmental biologists since the eggs are available in large numbers, are easy to handle and fertilize, but most importantly, they are transparent. One can, looking down the microscope, observe the behaviour of individual cells as the embryo develops. Because the cells move very slowly the process is best seen when speeded up in a time-lapse film. One frame of the film is exposed every six seconds and if the film is then shown at 24 frames a second, the motion is speeded up 144 times and provides a dramatic picture of the cells' behaviour. Because the embryos begin to swim at the blastula stage they must be trapped in a nylon net to keep them still, the square holes providing micro-aquaria through which sea water is slowly passed. After forty-eight hours the egg has developed into a free-swimming and feeding larva.

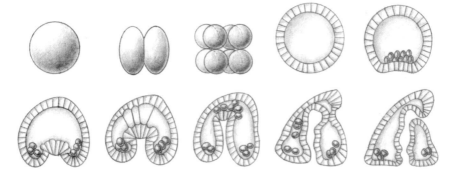

Following fertilization, the egg divides into two after an hour, and then there are further cleavages at half-hour intervals. The second cleavage is at right angles to the first and the third again at right angles. This regular pattern only continues for a short time, and after about eight hours, the embryo is a blastula comprising a hollow ball of about 1000 cells. The cells form a sheet just one cell thick enclosing a hollow interior. And then the blastula undergoes gastrulation. This involves a major change in shape, for the cells that will form the gut are a small patch on the outside and so must move inwards; this infolding of the wall will move right across the hollow interior to meet

the wall on the other side where the mouth will form. The process can be modelled by taking a balloon filled with water and pushing a finger in at one point until it touches the other side. The tube made by the indentation is the gut and the point where it starts will become the anus. The embryo has been transformed from a sphere into a torus; from a bun into a doughnut.

Gastrulation begins, not with the formation of the gut, but with the entry of the cells that will lay down the skeleton. In time-lapse films the first signs are pulsations by the cells where the gut will form. The cells seem to jostle for position and about 40 cells leave the wall and enter the hollow interior. These cells will later lay down the skeleton and move on the inner wall of the blastula to take up a characteristic ring-like pattern.

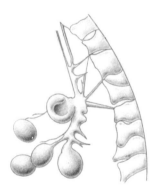

When these 40 or so cells have moved out of the wall they begin to put out fine extensions—very long thin filopodia—which can attach to the wall, contract, and so pull the cell towards the point of attachment. The force for contraction is provided by very fine muscle-like filaments within the filopod. Some filopodia simply fail to make contact with the wall and are withdrawn. Often, several make contact with the wall and then there is a tug-of-war between them. The winner seems to be determined by how strongly the extension adheres to the wall, and the cell eventually moves to the site where the attachments are strongest. It is not unlike climbing out of the sea on to a slippery rock. One comes out where one can get the best grip.

Sea-urchin gastrula held in a nylon net

By filopodial extension and contraction the cells migrate, and take up a ring-like pattern of the wall of the embryo. They do not go there directly. In fact, the pathways taken vary considerably from embryo to embryo. But gradually the pattern emerges and it can be seen that the cells' filopodia attach best in the ring region. The pattern of adhesiveness on the wall has a high point in the form of a band around the wall and so this is where the cells, through repeated explorations by their long filopodia, come to rest. This is where they make their most stable contact. With their fine processes they are always exploring their environments, even when they are in the band, finding the most stable contact regions. So it is the pattern of adhesiveness in the wall that determines where the cells go; and when, later on in development, this pattern changes, the cells follow—they seek stability. The wall thus provides a template for the pattern of the migrating cells.

Only after the entry of the skeleton-forming cells do the group of cells that will form the gut begin to pulsate on their inner surfaces and a small inward indentation of the wall occurs. This continues until there is a well-developed inward bulge which goes about half-way across the interior. Often there is no further change for an hour or so. But then the cells at the tip begin to shoot out long filopodia which make contact with the wall and contract pulling the future gut further in. The filopodia pull the sheet right across the cavity until it makes contact with the other side where it meets and fuses with another but much smaller invagination which is the future mouth. After fusion the cells in the centre of the zone of contact break down and there is now a channel—from an open mouth, through the tube of the gut, and out through the anus.

The filopodia at the tip of the gut guide the tip to the mouth region. When viewed in speeded-up films one sees a vigorous activity at the tip. Like the skeleton-forming cells, these cells send out many fine filopodia, some of which make contact with the wall. If they make stable contacts they pull the sheet to the site of contact, but others merely lose contact and are withdrawn. Since the future mouth region provides the most stable contact sites, the gut is pulled gradually to that region.

A DEVELOPMENTAL PROGRAMME

If the cells in the embryo 'know' where and when to change shape, contract, or move, then it begins to be possible to envisage a programme for the development of form. It becomes possible to think of accounting for all the changes in form of early sea-urchin development in terms of a changing pattern of cell contractions and cell contacts. We can think of this pattern of cell activities as being part of the embryo's developmental programme. It is a programme that does not describe the final form, but a generative programme that contains the instructions for making the shapes. A key feature of a generative programme is that it can be made up of quite simple instructions, yet generate very complex forms.

An analogy is origami—paper folding. The instructions for folding a piece of paper involve only a few operations like folding and unfolding. But the final form, a paper hat or a bird, can be very complex. It is very much easier to describe how to make the hat, than to describe the hat itself with its complex folds. The instructions provide a generative programme, the folding and unfolding are like the contractions and changes in contact in cell sheets in the embryo. Thus, what the egg contains is not a description of the adult but a set of instructions of how to make the animal and, using the same mechanisms, very different forms can be generated.

Contractions and changes in contact are only part of the programme and an essential component is the specification of where and when they occur—the process of pattern formation. But first I will look at some other changes in form, particularly those involving changes in shape in cell sheets, changes in contact, and cell migration. I am proceeding in the belief that there is a unity in developmental mechanisms, and whilst the sea-urchin provided an excellent model, there are other examples to be considered.

FOLDING SHEETS

Folding of cell sheets forms the basis of the early development of many structures as varied as the heart, lungs, brain, eye, and teeth. The early development of the vertebrate brain, known as neurulation, is not only important in its own right but provides a good model. Neurulation results in a major change in shape of part of the embryo—a flat sheet of cells on the upper surface folds up to form a tube that will develop into the brain and spinal cord. It is a process that can be directly observed in the embryos of amphibians and birds when gastrulation is completed.

In amphibia the first sign of neurulation is the flattening of the upper surface of the embryo into a broad plate bounded by two ridges or folds. At the anterior end of the embryo where the brain will form the folds are well separated with a broad area between them, but the folds run close together towards the rear end. Now, these folds

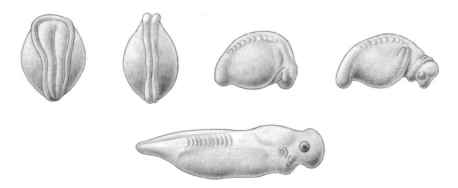

become more prominent, rise up, and approach each other, the sheet between them sinking down. The folds meet and fuse forming a broad tube, the neural tube, that will be the brain, with a narrower tube, to be the spinal cord, behind it. After fusion the tube sinks below the surface and is covered by a sheet of cells that will eventually form the skin. We will follow the later development of the brain and neural tube in Chapter 8. The failure of the neural tube to close is one of the causes of spina bifida.

The transformation of a flat sheet of cells into a tube during neurulation is mainly brought about by forces generated by cells within the tube, by the cells actively changing shape. On the upper surface of the sheet the cells contract, causing the sheet to bend, with the contractions on the concave side. Like the contracting filopods the contraction is due to the action of muscle-like microfilaments. When the sheet is examined in the electron microscope filaments are seen to be localized at the upper surface.

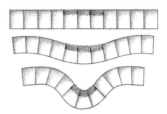

It has been possible to measure directly the forces that the cells generate during the formation of the neural tube. Minute dumb-bell-shaped pieces of metal were embedded in the two sides of the neural folds of an amphibian embryo and a magnetic field applied to prevent the folds coming together. The forces were found to be very small and use very little of the embryo's energy.

Development of the lens also involves bending of sheets. The eye has two different origins: the eyecup that forms the retina is an outgrowth from the brain, while the lens comes from the sheet of cells covering the embryo. At the same time as the eyecup approaches the sheet of cells covering the head region the lens begins to develop. (The co-ordination between the two will be discussed later.) The cells of the future lens actively elongate and then begin to fold inwards towards

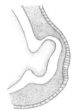 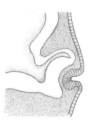 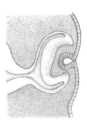 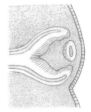

the eye, in a manner not unlike the early formation of the gut in sea-urchins. In this case the bending continues, now rather like neurulation, until an almost spherical sheet of cells protrudes inwards, and then detaches from the surface layer. This ball of cells will then develop into the lens.

This example of lens development shows how important it is to follow through the development of any particular organ. It is impossible by looking at any structure to work out how it developed. The only way is to actually see how it develops. There is no way, for example, that one could work out that the lens came originally from an overlying sheet of cells without following its development. It should be a condition, for anyone thinking about development to resist trying to infer how a structure develops by only looking at the final result.

Just as the formation of the neural tube and lens involves changes in the shape of a sheet of cells, so the formation of many other organs also arises from the folding and movement of cell sheets, which are caused by active change in the shape of the cells that make them up. Lungs initially develop from an outgrowth of the sheet that lines the gut near the mouth and then this outgrowth branches repeatedly to give the millions of microscopic lobes in the lung. The mammalian heart starts off as a straight tube and then bends, folds, and, together with further growth and subdivision, gives the four chambers that pump the blood. Even the shape of our teeth is largely determined by the foldings of the sheet of cells that will lay down the enamel.

Many human developmental abnormalities arise from abnormalities in the moulding of sheets of cells. Sheets fail to separate, or close, or fold properly. As spina bifida can result from incomplete closure of the neural tube, so abnormalities in the development of the gut can be due to failure of tubes to seal properly. Again certain kidney abnormalities arise because cell sheets that form tubes do not connect up correctly.

It is possible to make computer models that simulate many of these changes in shape, the model being based on the asymmetric contraction of microfilaments. Imagine a hollow tube made up of a layer of cells just one cell thick with contractile filaments located near the outer surface. Imagine now that from a point—say the top—contraction is

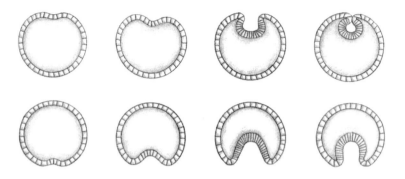

initiated that spreads. Depending on how fast the cells contract and how far contracting cells extend, a variety of different shapes result. In one case the sheet folds in a way that is very similar to the formation of a neural tube; in another it is very similar to the way the lens forms; yet in another it is like the early formation of the gut in sea-urchin development. Even small changes in how fast and how far the contraction spreads can have profound effects on form. How these processes could be controlled, particularly by the genes, is a central problem.

REPEATED STRUCTURES

The famous molecular biologist, Francis Crick, once remarked that embryos seemed to be very fond of stripes. What he meant was that there are numerous examples of repeating patterns, and embryos seem to like breaking up into repeated units. One has only to glance at a human skeleton to see the numerous segments of the vertebral column. Each vertebra may be different from its neighbour but is, clearly, also very similar. Repeated segments are also found in insects, worms, crustacea, and many other animals. If one looks at the early development of vertebrae it becomes clearer just how segmental much of our own early development is.

In vertebrates, shortly after gastrulation, the brain can be seen forming at the anterior end of the embryo. Behind it, stretching backwards, like two lines of paving stones, are the somites. The somites are blocks of tissue that will form the vertebrae and the muscles of the

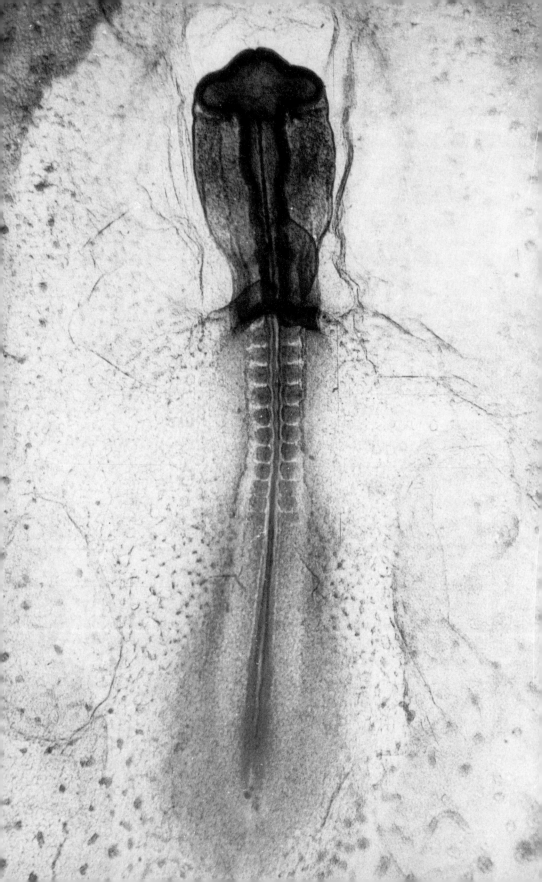

back. They develop, pair by pair, lying either side of the future spinal cord. First, just two appear at the front end as distinct blocks of tissue and then, about each hour, another pair are added behind them and a wave of formation proceeds backwards so that at the end of a few days there are 46 somites. The first few pairs of somites that form do not persist; they disappear and are incorporated into the head. This is thought to reflect our evolutionary history in that as the head became larger, the development of the anterior somites was modified.

The mechanism whereby the somites develop is still not well understood. What is clear is that the mechanics of the process, that is, the breaking down of the tissue into discrete blocks, involves an increase in adhesion between the cells in each somite. But what specifies just which cells will form an adhesive block separating each somite from its neighbours is not known. However, the sequence of somite formation and what they will form seems to be programmed early in development long before the somites themselves appear and provides a good example of how, once the programme is specified, the cells follow it even when they are placed in abnormal relationships.

If a strip of the tissue that has yet to segment into somites is cut out of a chick embryo, turned through 180 degrees and replaced, the somite formation proceeds normally up to the site of the graft, but will then continue from the rear edge of the inverted piece, the sequence now going in a direction opposite from normal until the operated piece is fully segmented, and will then continue normally, again from the rear edge. It is as if there were a built-in timetable as to when somites are formed. The wave of somite formation that is observed progressing from the front to the back of the embryo thus does not involve any propagation of a message. It merely reflects the implementation of a programme laid down much earlier.

CELL AFFINITIES

The importance of cell adhesion in moulding form has already been referred to. During gastrulation, differences in cell adhesion guide the cells; during lens formation the lens cells lose adhesion with adjacent

Somites in a two-day chick embryo

cells at the time of detachment; and somite formation is essentially an increase in adhesiveness locally. Changes in adhesiveness are an essential part of the developmental programme.

The molecular basis of adhesiveness lies in cell adhesion molecules—CAMs—at the cell surface. CAMs allow cells to bind to each other by a mechanism in which the molecules of the CAMs show a specific affinity for each other. These molecules are proteins, which are embedded in the cell's surface membrane and have one portion that sticks out and binds to a similar or complementary molecule on adjacent cells. Part of the binding mechanism may involve ions like calcium, so when calcium is removed, the cells of some early embryos just separate into individual cells, and restoring the calcium causes the cells to adhere again. The crucial point about CAMs is that they can provide specific adhesion between cells, and that cells express different CAMs at different stages in development. For example, at the time when the neural plate starts to fold up, the CAMs on the surface of the embryo are all the same. As the tube folds, the CAMs of the future nervous system change and become different from those of adjacent cells. This allows fusion and separation of the tube. Similarly, the future lens cells cannot detach from the sheet unless their CAMs change to allow them to let go of the neighbouring cells.

Molecules like CAMs at the cell surface can provide the basis for self-assembly of some organisms, in a manner that is analogous to crystal growth. The molecules which form the crystals, under the right conditions, assemble spontaneously and the form of the crystal arises directly from the property of the molecules. The crystals of salt look different from those of, say, sugar because of the way the molecules assemble. The key point about self-assembly is that the forms generated derive from the nature of the elements that make them up. In a similar manner, if cells of the sponge (the bath sponge is the skeleton laid down by the sponge) are separated into a random mixture of individual cells they will actively move around and become reorganized into a normal sponge, with the cells in the correct relationship with one another. It can be explained largely by adhesive affinities leading to self-assembly. The crucial nature of the adhesive surfaces of the cells is again illustrated if the cells of two different species of sponge are mixed together; they sort out to give the two normal sponges. The cells will not adhere to cells of a different species.

It is not just sponges that can self-assemble. The small freshwater hydroid, hydra, has a glove-like form, its tentacles being used to capture prey. (Hydra has remarkable powers of regeneration, see Chapter 13.) Hydra, too, can be separated into its individual cells, which will reassemble to form more hydra. Even the early sea-urchin embryo can be separated into single cells and will reform a more or less normal embryo. But with higher animals this does not occur. One cannot separate a frog into its individual cells, mix them together, and get a frog; the results will be a chaotic mixture.

One can, however, disaggregate some embryonic organs into cells which will reaggregate into patterns resembling the tissue from which they come. Cells generally associate with those of their own class. They seem to enjoy contact with like cells. So, if groups of cells are removed from the early amphibian embryo, separated into individual cells, and then reaggregated, the cells mill about and sort themselves out so that, for example, cells that are normally on the outside of the embryo, such as the future skin cells, surround those that are normally on the inside, like future muscle cells. Or if liver and limb bud cells are mixed they too sort out into large clumps; but they make neither a proper liver nor limbs.

MIGRATING CELLS

When cell sheets are deformed and folded the cells tend to keep contact with their neighbour and the sheet behaves as a coherent social group of cells. But cells may also move as individuals over quite long distances, such as the skeleton-forming cells in sea-urchins. In vertebrates, there are two groups of cells that show a remarkable capacity for migration—the neural crest cells and the nerve cells.

Neural crest cells are a group that arise at the site where the folds of the neural tube fuse. At the site of fusion, cells leave the sheet and form two clumps of cells on either side of the neural tube. These cells

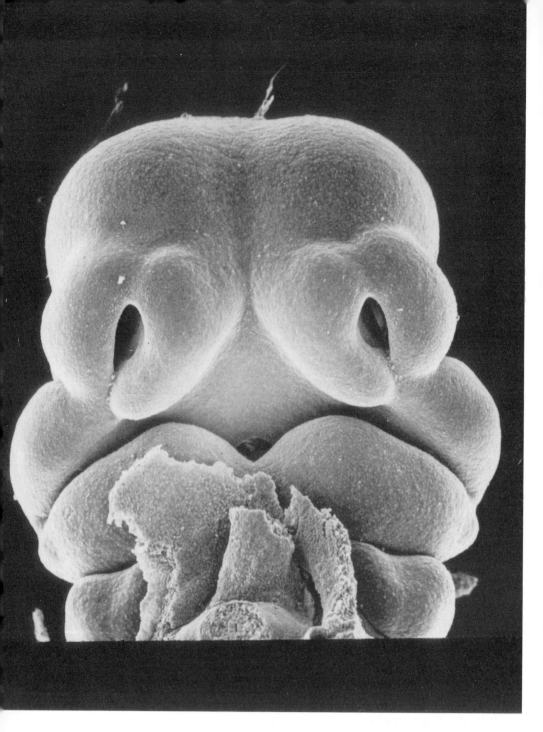

Embryonic face

then migrate to many different parts of the embryo. Some cells migrate beneath the future skin and will give rise to pigment cells. Other cells give rise to sensory nerves, to nerves in the gut, to cells in glands like the adrenal, and to cells that provide the insulation for nerves. Yet other neural crest cells migrate into the head region and form tissues of the head and face such as cartilage and bone. If insufficient cells migrate into the front of the head the face will be abnormally small. The pathways taken by the cells may be determined by the nature of the extracellular matrix and contacts with other cells *en route*.

Probably the most impressive migration of cells is that of nerve cells—neurons. These have special features and are described in Chapter 8.

Another dramatic migration is that performed by the primordial germ cells (Chapter 9). These cells have their origin a long distance from the sex organs. Because they are larger than other cells and have special staining cytoplasm that is easily recognizable, a trail of migrating cells can be seen in the embryo. The cells seem to migrate along a track of extracellular material that covers cells which are oriented along the pathway they take.

It may be wondered if gradients in some chemical might guide cells to their destination. There is no doubt that some cells can exhibit chemotaxis, that is, move towards the source of chemical which is diffusing in the medium. Certain white blood cells do this in response to an infective agent. The cellular slime moulds, a primitive organism related to fungi, aggregate in response to a chemical gradient. But for most developmental systems, there is, perhaps disappointingly, little evidence for such a mechanism.

GROWING A FACE

Growth is an important mechanism for generating changes in form. In general, growth is a moulding process, and occurs at later stages of development when the main body plan has already been laid down. For example, our face and how we look very much depends on the relative growth of the different parts. At an early stage in development the human face is made up of a series of bumps, which looks quite grotesque, and there is little hint of the face that will emerge. The bumps, or processes, as they are called, contain many cells covered by a cell sheet. There are two processes, one on either side that will develop into the lower jaw; two further processes that will form the cheeks and upper jaw, and a central process that will give rise to the nose. Each process has its own characteristic growth pattern and together these generate the face. For example, the little depression in the centre of the upper lip is where two of the processes meet. One can think of the differences of being handsome, beautiful, or just ordinary, in terms of small differences in the growth programme for each region. Beauty can be just one more cell multiplication away.

FORM AND PATTERN

Contractions, changes in adhesion, cell movement, and growth are all cellular activities that go to mould the form of the embryo. These are by no means the only mechanisms but they are amongst the most important. The mechanisms for generating form are in some cases poorly understood. Even so, we can see how form can emerge during development from the varying patterns, both in space and time, of such cellular activities.

If we can understand the cellular forces that bring about changes in the shape of the embryo then we can begin to ask further questions. Why do particular cells exert these forces and not others? Why do some groups of cells grow while others do not? These are really questions about the organization of the spatial organization of cellular activities. To follow the analogy with origami, we need to know where to fold the paper. The same set of cell activities are used again and again—contraction, movement, change in adhesion—and what makes organs different is how these activities are organized in space and time. That is the problem of pattern formation.

PATTERN FORMATION

COMPARE ONE'S body to that of a chimpanzee—there are many similarities. Look for example, at its arms and legs, which have rather different proportions from our own, but are basically the same. If we look at the internal organs there is not much to distinguish a chimpanzee's heart or liver from our own. Even if we examine the cells in these organs we will again find that they are again similar to ours. Yet, we are different, *very* different from chimpanzees. Perhaps you may wish to argue, the differences lie within the brain. Perhaps there are special brain cells which we possess that chimpanzees do not. This is not so. We possess no cell types that the chimpanzee does not, nor does the chimpanzee have any cell types that we do not have. The difference between us and chimpanzees lies in the spatial organization of the cells.

Now compare your arm with your leg. They contain the same cell types—muscle, tendon, skin, bone, and so on—yet they are different. Again, the explanation lies in how these cell types are spatially ordered. The principle of different spatial patterning accounting for the differences in animals applies right across the vertebrates. While there are some differences in the cell types that make up fish, frogs, birds, and ourselves, the main difference lies in the spatial organization of the cells.

All vertebrates have basically the same building blocks but they are put together in different ways. How this is done is the process of pattern formation.

Patterning poses the problem of how the cells know what to do, how to behave. In the moulding of form we saw that changes in shape of the embryo could be accounted for in terms of localized contractions and changes in cell adhesion. How are the cells that will change shape specified to do so rather than all the other cells? As in the origami analogy (Chapter 2), what specifies where the paper will be folded? To use yet another metaphor, moulding of form can be thought of as metalworking; patterning like painting.

A HARMONIOUS SYSTEM

Belief made them see it. In the seventeenth century some preformationists claimed they could detect in the head of the sperm cell a tiny person in miniature—a homunculus—just waiting to emerge. Other preformationists thought the adult was present in miniature in the

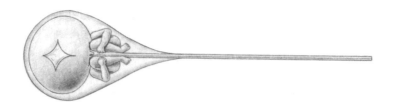

egg. Although cruder versions of such theories had been abandoned by the end of the nineteenth century, preformationist ideas still present an important theory. The question is to what extent the pattern of development is already present in the egg. To what extent is the pattern already preformed? Could it not be that during cleavage, when the egg divides up, each cell acquires special determinants, either cytoplasmic or nuclear, which would control the cell's future development? With further cell divisions there would be a further unequal distribution of such determinants. In essence, this was the preformationist theory proposed in the 1890s by August Weismann, one of the uncles of modern embryology. He drew an analogy with an army: the nucleus of the fertilized egg contained the whole army and as the egg cleaved, so different brigades, like the muscle brigade, or the cartilage brigade, would be distributed to different cells. Although he helped make development a causal science his theories were largely wrong; even so, his preformationist ideas, in which nuclear determinants were unequally distributed during cleavage did stimulate the key experiments.

Another German biologist, Wilhelm Roux, has a better claim to being the father of our subject. He coined the term 'developmental mechanics' and was one of the first to attempt a causal analysis of early development. With a hot needle, he killed one of the two cells of the frog embryo after the first cleavage and then watched the development of the surviving cell. A typical half embryo was seen to emerge just as if an older embryo had been sliced in two with a razor. It seemed that the detailed pattern of the embryo was laid down in the egg and became partitioned during cleavage, thus supporting Weissman's claim. But his fellow countryman, Hans Driesch, was not persuaded:

> Roux's results were published for the first time in 1888: three years later I tried to repeat this fundamental experiment on another subject and by a somewhat different method. It was known from the cytological researches of the brothers Hertwig and Boveri that the eggs of the common sea urchin are able to stand well all sorts of rough treatment, and that, in particular, when broken into pieces by shaking their fragments, will survive and continue to cleave. I took advantage of these facts for my purposes. I shook the embryos rather violently during their two-cell stage, and in several instances, I succeeded in killing one of the cells, while the other one was not damaged . . . Let us now follow the development of the surviving cell. It went through cleavage just as it would have done in contact with its sister-cell . . . So far there was no divergence from Roux's results . . . and indeed the next morning a *whole* diminutive blastula was swimming about. I was so much convinced that I should get Roux's result in all its features, that even in spite of the whole blastula, I now expected the next morning would reveal to me the half-organization of my subject once more; the

gut, I supposed, might come out quite on one side of it, as a half-tube, and the ring of cells might be a half one also.

But things turned out as they were bound to do and not as I expected; there was a typically *whole* gastrula on my dish the next morning, differing only by its small size from a normal one; and this *small but whole* gastrula developed into a whole and typical larva . . .

That was just the opposite of Roux's result and was the first clear demonstration of the process known as regulation: the ability of the embryo to develop normally even when some portions are removed or rearranged. Driesch provided further examples of regulation, for, even at the four-cell stage each cell could develop into a perfectly normal embryo. This showed quite unequivocally that the pattern was not laid down in the egg. For if all the parts were neatly laid down in the egg like a little homunculus, then separating the embryo into two or more separate parts would divide the pattern in two or more and each should form just a fragment of the whole. Clearly this was not the case.

Driesch went on to isolate other fragments of the sea-urchin embryo and he claimed that each fragment, or combination of fragments, no matter how they were arranged, gave rise to a small but normal embryo. He concluded, in the language of the time, that the early sea-urchin was a 'harmonious equipotential system' in the sense that the parts all functioned so as to generate a normal organism. More specifically, his experiments showed, he claimed, that the cells always developed according to their relative position within the embryo, and this sense of position required that there be a self-organizing coordinate system which told the cells their position and that they then knew what to do. Such self-organization of a coordinate system was, he claimed, quite beyond conventional science, and required a special biological force—entelechy. The essence of his argument was that life in general, and development in particular, could not be encompassed or explained in terms of physics and chemistry. He was a Vitalist, searching in vain for a mysterious life force.

Driesch was wrong for two reasons. First, isolated parts of embryos do not always develop normally. He was perfectly well aware of this but chose to ignore it. Secondly, as I will show, it is quite easy to imagine a self-organizing system which specifies the relative position of cells.

Driesch chose to ignore experiments in which parts of the embryo do not develop normally. He was quite right to claim that each of the four cells at the end of the cleavage would form a normal larva. But if at the eight-cell stage the embryo was separated along the plane of the third cleavage into two groups of four cells, then each of these two fragments developed quite differently. One developed into nothing more than a simple ball of cells with no gut at all, the other into a more or less normal larva. So much for 'harmonious equipotentiality'.

We now know that the sea-urchin egg has a well-defined polarity from the beginning and that there are cytoplasmic differences along the egg. The site where it is attached to the ovary is known as the animal pole and the other end the vegetal pole. The vegetal pole is the region where the cells of the gut will develop and where the group of skeleton-forming cells will enter. The animal and vegetal poles define the main axis of the embryo and the first two cleavages are parallel to it while the third is always at right angles to it. Experiments separating

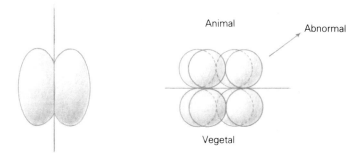

the cells at the two- and four-cell stages give cells with a complete animal vegetal axis and each, in isolation, develops normally. However, if the embryo is divided into two along the animal–vegetal axis the two halves do not develop normally. This is most easily done after the third cleavage which is at right angles to the first two, and so divides the embryo into four animal and four vegetal cells. The animal half, if isolated, will only develop into a simple hollow sphere. Thus, there is something in the cytoplasm of the egg, from the very beginning, which makes the animal region different from the vegetal. In some species there is even a row of coloured granules near the vegetal pole which are a visible marker of this polarity.

Even though Driesch was wrong in assigning limitless regulative properties to the early sea-urchin embryo, he did demonstrate that the embryo had a remarkable capacity for regulation. Considerable experimental rearrangement of the cells can still result in normal development. We are thus faced with the following problem: what organization in the early sea-urchin embryo ensures that the proportions of cells that form the skeleton, the gut, and the surface layers, will be the same in embryos whose cells can be quite extensively rearranged and whose size varies over a factor of eight-fold? From quarter embryos to giant embryos, formed by fusing two eggs together, normal larvae emerge.

Yet for the early mammalian embryo, it seems that Driesch was correct for there is no polarity laid down in the egg. In the mouse egg and at least up to the 16-cell stage all the cells seem equivalent with no fixed fate. It is possible to rearrange the cells of the early mouse embryo in numerous combinations and normal development will still occur. Even if several mouse embryos are pushed together so that they fuse and this large mass transferred back into a mother a normal mouse will still develop.

In humans, the development of identical twins or more dramatically, quintuplets, again illustrates the absence of a fixed pattern in the egg. Surprisingly, identical twins rarely arise from the separation into two cells at the two-cell stage. Instead, the separation occurs much later when the embryo is made up already of many hundreds of cells. This means that in human embryos even when there are several hundred cells present the fate of the cells is not fixed and if divided into two, two normal embryos can still develop.

THE FRENCH FLAG PROBLEM

How then do the cells 'know' to make the right pattern? It is quite convenient to simplify the problem, and instead of thinking about sea-urchin or mouse embryos, to think about flags. Consider the following problem. Imagine there is a line of cells, each of which can turn blue, or white, or red. The problem is to generate a pattern that looks like a French flag, that is the first third of the line of cells is blue, the middle third is white, and the last third is red. What organizing principles must the cells have so that they reliably give this pattern, even when the cells are rearranged and the length of the line varies? (It is not unhelpful to imagine oneself standing in such a line: how would one decide whether to be blue, white, or red?)

There are a number of solutions, but the most interesting solution is for the cells to 'know' their *position* in the line. If they know their position with respect to the ends, then they can 'work out' which third they are in and so decide whether to be blue, white, or red. They could do this by 'numbering' from the two ends and so know their position in the line. It is then quite easy to work out which third the cell is in, and so whether to be blue, white, or red. Moreover, if the cells keep monitoring their positions it will not matter if, at an early stage, some of the cells are removed from the line. We can begin to see how the system would regulate back to normal when parts are removed. Even if flags are not embryos there are many cases in development where the embryo does behave rather like a regulating French flag.

There are, then, two quite difficult things that cells have to do. The first is to know where they are, that is, acquire positional information, and the second is to use this information appropriately. We know much more about the former than the latter. But given that they can have their position specified, and can interpret this position so as to form the pattern, then this provides a very powerful means of generating patterns. For example, if we extend the pattern to two dimensions then, again, just using cells that can become blue, white, or red, we can have a French flag, or a Stars and Stripes, or the Union Jack. The method provides a way of generating very complex patterns.

Imagine the cells as if they were in a spectator stand. Each cell would
have row number and seat number and so be uniquely identified. In
addition, each cell must have a set of instructions—possibly analogous
to genetic information—which lists what every cell must do in every
position. The cells just look up their position in this set of instructions
and behave accordingly. It may be, as we will see later, that for each
pattern the cells have the same positional information but just have
different rules for interpreting what to do. The rules for interpretation
will depend on the genetic constitution of the cells and their developmental history.

If cells are to have their position specified as in a coordinate sytem
then it is essential to have a boundary region or origin, from which
position is measured. It is also essential to have some way of measuring distance from this boundary region. One way would be to have a
chemical whose concentration was fixed at one end of the line, and this
concentration decreased as one moved down the line. This would provide a gradient in the concentration of the chemical and if the cells

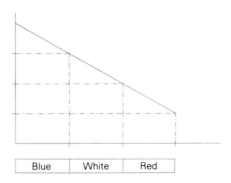

| Blue | White | Red |

could read the concentration they would know their position in the
line, with respect to the boundary. Using the concentration of chemicals is just one of several possible ways of specifying position. It may
well be that the main cross-talk between cells involves specifying position. If we could listen to the conversation between cells talking to
each other during development it might be anything but fascinating. If
all the talk is about position all we might hear would be equivalent to

'You be number 33', 'OK, you be 32 and I'll tell my neighbour to be 34'. Not a very interesting conversation but crucial for patterning.

Communication between cells takes place in a limited number of ways. The obvious one is by a chemical message leaving one cell and diffusing to the next—rather like speaking. If the chemical diffuses a long way it is more like shouting and a larger number of cells could receive the signal. Another type of communication requires direct contact between the membranes of cells because fine pores, known as gap junctions, develop at the site of contact and then allow direct communication between the cytoplasm of the two cells. The signal can only be a small molecule as the pores are so fine as to prevent passage of larger substances. These channels have the advantage that the signal remains within the cell membrane and never enters the external medium—rather like a private line. One further possibility is that the extracellular material provides a means of communication.

If cells have their position specified as in a coordinate system and also have rules for interpreting the positional information, then it is possible, with the appropriate rules for interpretation, to generate any pattern that is required, from faces—even the Mona Lisa—to limbs. The extent to which cells actually use such mechanisms is still being investigated.

How big are positional fields? Over what distance do cells have to communicate in order to establish their position? Remarkably, the distances are very small. In the sea-urchin, positional signals would not have to travel across more than 20 cells. No system has been described in which positional signals have to be transmitted over more than about half a millimetre, or about 30–50 cells. The small size of positional fields has two important implications. First, simple diffusion of chemicals could provide the signals; secondly, if patterns are laid down in such small fields, their later development may largely be due to programmed growth. Embryos are organized so that inter-actions between cells occur on a small scale.

Returning to the sea-urchin embryo, it can be seen that the animal–vegetal differences could be used to establish the boundary regions, that is, which end is to be which. The cell-to-cell interactions could provide the cells with positional information. A classical experiment illustrates this possibility. The animal half of the early embryo develops as a simple ball of cells—but if it is combined with cells from the vegetal pole a normal, but smaller embryo will develop. It is as if the vegetal cells have set up a new boundary region for specifying positional information.

EGGS AND AXES

Eggs are spherical but give rise to animals with well-defined axes with heads at one end and 'tails' at the other. There are in almost all animals two main axes, the antero-posterior which defines the 'head' and 'tail' ends and the dorso-ventral axis which is at right angles to it. Our faces are ventral, the back of our heads are dorsal; one sits on the dorsal surface of a horse. How are these axes specified in development?

In some eggs like the sea-urchin and the frog egg there is, from the beginning, a well-defined polarity in the egg which defines an axis with respect to which the antero-posterior axis and dorso-ventral axis will be specified, but in a manner which is not straightforward. The antero-posterior axis of the frog egg more or less corresponds to the animal–vegetal axis. But the dorso-ventral axis is specified by the point where the sperm, which fertilizes the egg, enters. That site defines the ventral side. For the mouse egg there are no axes and both the antero-posterior and dorso-ventral are specified, in a manner which is not yet understood, during development.

REGULATION AND FATE

It is a fundamental belief of developmental biologists that the regulatory processes, just discussed, are central to how the pattern is established in normal development, and are not just a bizarre result of experimental interference. But does regulation continue throughout development, or is it just a property of the early embryo?

It is possible to label the cells of the early embryo and follow what they do during development. This provides a way of finding out their normal fate. The cells in one region of the embryo can be marked with a dye, or one can inject, with a very fine hollow tube, a fluorescent dye into a single cell so that a specific cell and all its offspring can then be followed through development. In this way it has been shown, for example, that the muscles of the newt always come from a particular region of the early embryo, and the eye from quite a different region, and the gut from another. The origin of all organs can thus be mapped and so a 'fate map' can be plotted which indicates to what the normal embryo will give rise. The gut comes from the yolky part of the early embryo, the skeleton and muscles from the middle region, and the nervous system from the upper part.

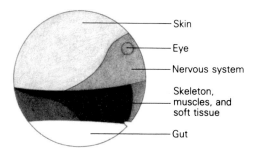

In some ways this fate map might present a puzzle. How could it be that the cells that will form the gut and the bones and muscles are on the outside of the early embryo, yet they will form structures that are clearly on the inside of the animal? The answer, in part, lies in gastrulation. The cells from the outside move into the interior during gastrulation and only at the end of the gastrulation are they more or less in their proper position. It is during gastrulation that the main body plan is laid down (Chapter 2).

The fate map should be regarded rather like a train timetable—it tells you only what will normally happen. It in no way means that other outcomes are not possible, particularly if the system is perturbed by say, bad weather, or a strike, or in the case of embryos, by experimental manipulation. If cells from the region of the early embryo that will normally give rise to the eye are grafted into the region that will form the gut the cells do not form an eye any more but just part of the gut. This is just another example of regulation but it also shows that in the early frog embryo the fate of the parts is not fixed. In general, if cells of vertebrate embryos are moved from one part to another of the early embryo they develop according to their new location and not from where they are taken. Their fate is dependent on their new position in the embryo: they respond to their new address.

This flexibility of fate does not last for long. The fate of the cells becomes, with time, more and more restricted until it is effectively fixed. So, if at a later stage in development, such as after gastrulation, the region that will form the eye is grafted to the belly region of the frog embryo, on this occasion it continues to develop there as an eye. The result is the development of an eye, isolated and unseeing, in the belly of the embryo. This result is quite a general one. Whereas early on cells develop according to their position in the embryo, with time their fate becomes fixed. When they are grafted to a foreign site, they continue to develop as they would have done at the site of origin. The cells, with time, acquire an autonomous developmental programme and no longer respond to new positional cues.

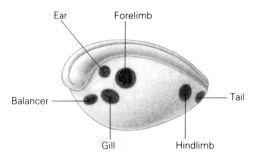

The process by which cells have their fate fixed is known as determination. An almost universal feature of determination is that it involves subtle chemical changes, almost certainly turning on or off genes, and the overt result may not be seen for many hours. When the presumptive eye is grafted, after gastrulation, to the belly region and develops into an eye, the tissue shows no outward sign that it will develop into an eye. Thus, by the end of gastrulation, one can draw a map on the embryo showing the developmental path to which various regions are now committed. The embryo has been broken up into a number of regions whose development is largely independent of one another.

Positional signals are only one kind of communication between cells. Of equal importance are signals between groups of cells, a process known as induction, which historically at least, is considered to be the major mode of interaction in the development of embryos.

INDUCTION

There has been only one Nobel Prize in embryology. It was awarded to Hans Spemann, yet another German embryologist, in 1935 for discovering the organizer. In 1924, he and his collaborator, Hilde Mangold, grafted a piece of tissue from a region in the early newt gastrula where cells from the outside are just beginning to stream inside, to the opposite side of another embryo. The result was astonishing—a whole second embryo developed in the region of the

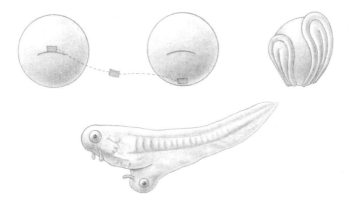

graft. The graft had induced the host tissues adjacent to it to completely change their fate and form a second embryo in relation to the graft, and for this reason it was called the organizer. It was as if the graft was setting up a whole new embryo and re-specifying the positions of the cells in its vicinity. A further very important feature of this type of experiment was the demonstration that the nervous system will only develop if future muscle and adjacent cells on the outside move in underneath the outer layer. These migrating sheets of cells produce a signal which induces a nervous system in the overlying sheet.

The discovery of induction had a profound influence on experimental embryology. It was, foremost, the first unequivocal demonstration of cell-to-cell interactions. While the idea of such interactions could be inferred from earlier experiments, here the result was much more dramatic and clear-cut. One tissue—the organizer—could so influence other tissues with which it was placed in contact that a whole new embryo could be formed. Another important feature lay in the way the experiment had been done. In order to make sure that the grafted tissue really induced a new embryo in the host, it was necessary to have markers that would enable graft and host cells to be distinguished. Spemann used the natural difference between two species of newt to this end, one being pigmented, the other not. The graft was from the unpigmented species into a pigmented embryo, and the new axis was clearly pigmented.

Spemann and Mangold's discovery of the organizer did not come as a surprise to Spemann, nor did it involve any element of luck. It was the result of a carefully planned experiment based on numerous other experiments performed over a long period. It is possible that the experiment had been done by an American embryologist many years earlier but its significance was overlooked. Spemann is supposed to have said 'A discovery can be unexpected, but not un-noticed'.

Embryonic induction seemed to provide an excellent system for further analysis. It led, for example, to the quest for the nature of the signal from the organizer that caused the second embryo and the second nervous system to form. Hopes were high when it was found that even if the organizer tissue were killed it could still induce, for this offered the possibility of extracting the responsible chemical signal. Alas, this turned out to be much more difficult. All sorts of substances and tissues could act as inducers; from rat liver to heat-killed neural tissue itself. Too many substances had positive effects. In spite of thousands of experiments the nature of the signal to this day remains unknown. One case where the inductive signals may have been identified is in early amphibian development.

In the fertilized amphibian eggs, the upper animal region is heavily pigmented whereas the lower, vegetal, half is white and heavy with yolk. The pigmented animal portion—the animal cap—will develop into skin and nervous system, the yolky vegetal portion will form the gut, and the middle portion the skeleton, muscle, and soft tissues. These are the normal fates, but for muscles at least, that pathway is not specified from the beginning but requires signals from the lower vegetal position. If the future muscle-forming region is isolated from the rest of the embryo at an early stage and cultured it will not develop into muscle. For muscle development an inductive signal must pass from the vegetal region.

The standard experiment to demonstrate the inductive signal from the vegetal region makes use of the animal cap. An isolated animal cap will never form muscle and will only form a simple sheet of cells. But if the animal cap is combined with the vegetal region some of its cells will now make muscle. Recently, the inducing chemical agents from the vegetal region have been identified and if added at very low doses to isolated animal caps, muscle will develop. These inducing molecules, the first to be characterized, are proteins that are, surprisingly, identical to growth factors (Chapter 11) that operate at late stages in development and which also act as signal molecules in the adult. One of these molecules, activin, is involved in the hormonal control of reproduction. So the same signals can be used for quite different purposes. It may even turn out that these growth factors are the signals from the organizer which specify the main body axis.

This experimental system has also illuminated the social nature of cells, since their response to the inducing signal in early amphibian development is enhanced by being in a crowd. The animal cells' ability to respond to induction by the vegetal region was tested by placing them in a sandwich between two vegetal regions. If the cells were placed as a solid clump in the sandwich, induction was good and many of the cells developed into muscle; but if the cells were placed in the sandwich as a thin single layer no muscle cells developed. In order for induction to occur the cells need to be in a community of sufficient size, a phenomenon termed 'community effect'; the cells' development being dependent on the presence of other similar cells.

There are many other examples in development of inductive interactions, one tissue signalling to an adjacent tissue so that its developmental pathway is altered. Induction thus provides a powerful means of co-ordinating the behaviour of adjacent groups of cells both early and late in development. For example, the eye is a beautifully organized and complex structure but the parts of which it is composed have, as we have seen, very different developmental histories. The main part of the eye, the retina, is an outgrowth from the brain, whereas the lens develops as an invagination from an overlying sheet of cells. The lens forms at precisely the spot where the outgrowth of the eye, in the form of an eyecup, approaches the surface. This co-ordination of two developmental events, crucial to the formation of the eye, is achieved in some animals by the approaching eyecup inducing the lens to develop just where it touches the surface, and so the lens develops at the right place. By grafting in another eyecup another lens can be induced to form from the overlying layer. Again, the nature of the signal is unknown.

Many cases of induction involve a signal from the cells underlying a cell sheet. One of the most dramatic examples is the development of the enamel that covers teeth. Teeth, like eyes, develop from two groups of cells. There is a sheet of cells covering the tooth germ which is made up of a loose mass of cells. The enamel comes from the cell sheet and is induced by the cells of the underlying tooth germ. That is not dramatic. The drama comes from separating the tooth germ of a mouse embryo from the overlying sheet and then recombining it with the cell sheet enclosing the limb, but which would normally form skin. This new construction can be cultured and develops quite a nice tooth, but now the sheet of cells that would have become the skin of the foot is induced to make enamel.

Whatever the nature of the inducing signal, it is a signal that can be recognized by quite different species. An old and elegant result, at first puzzling, involves combining the surface sheet of cells of a frog embryo with the underlying tissues from near the mouth region of the newt. Frog embryos develop into tadpoles with suckers near their mouths with which they attach to rocks or plants. Newts never develop such structures. Yet, when frog surface layer is placed on the newt near the mouth region, suckers develop in just the right place even though, in evolution, newts and frogs separated millions of years ago. Perhaps the newt is just providing the frog surface layer with positional signals that may still be the same in both animals: what has changed in evolution is how the signals are interpreted.

INSTRUCTION OR SELECTION

Do signals really instruct cells as to what to do? Consider a juke-box: if you select one of 20 tunes, have you really given the system an instruction or have you merely selected one of the 20 tunes from the repertoire of the machine? Instructions imply providing new information to the recipient. Are, then, cell signals instructive in that they tell the cell something it does not already know? Many signals are merely STOP or GO. STOP signals prevent the cell progressing on its developmental pathway and GO signals can only direct it along new pathways, if these are open to the cell. At any stage, the options open to a cell are very small in number, often only one or two. For example, the surface sheet can develop a variety of surface structures depending on the underlying inducer but it cannot develop into any of the embryo's internal structures, like muscle or cartilage. Positional signals might be thought of as instructive since they tell the cell its position in the system. But the cell must already have an internal system which can respond to such signals and again the signal is being selective. All signals are thus essentially selective rather than instructive since they bring a minimal amount of new information to the cell, and only select from the possible responses open to the cell. This emphasizes the importance of the cell's internal programme.

Once it is realized that signals that change cells' behaviour do not really carry intricate information, then it can be seen that any complexity of behaviour lies in the cells' capacity to respond rather than in complexity in the signals. Since many signals could be of the

STOP and GO type, complexity would emerge because the cells' responses would be determined by their past history. The current cell state determines the response rather than the particular signal. This is analogous to the complexity of computers where the signals are essentially noughts and ones but the response depends on previous events.

In emphasizing simplicity of signals I may have gone too far. There are very few cases where all the signals between cells are known. It could well be that there are many more signals operating between cells than is currently suspected. Nevertheless, my own belief is that this is not the case, but only further research will tell. For the moment I cling to the idea that there are a relatively small number of cell-to-cell signals.

A SIMPLE PATTERN: THE EARLY MOUSE EMBRYO

The development of the mouse embryo is the best experimental model for human development. It is possible to culture the early embryo and then return it to the mother's uterus where it will develop normally. Cell divisions of the mammalian egg are unlike those of lower animals where the planes of division appear with military precision. The early divisions of the mouse embryo seem much more sloppy and between the 8- and 16-cell stages the surface of the clump of cells becomes

smooth and at the 32-cell stage there is a single layer of cells enclosing some cells on the inside. The fate of these two groups of cells, those on the outside and those on the inside, is quite different. Those on the outside form the trophoblast which does not give rise to any structures in the embryo proper but is involved in the implantation of the embryo in the uterus and the formation of the placenta. The embryo proper comes from some of the cells on the inside—the inner cell mass—and goes on to gastrulate. Thus, quite a lot of early mammalian development is devoted to setting aside most of the cells for extra-embryonic structures rather than the embryo itself. It is in just those few cells in the inner cell mass that we have our origins.

The patterning problem here is what specifies which cells will form the trophoblast and which cells will form the inner cell mass. It is certainly not due to anything laid down in the egg or due to anything special about the first two divisions. There are numerous experiments in which the cells of the four- to eight-cell stage embryo are rearranged and recombined, always giving rise to normal development. The most likely explanation is that the patterning results from the difference in the position of the cells and hence their different environment. Those cells that end up on the inside become inner cell mass and those that end up on the outside, the trophoblast. It is thus partly a matter of chance which cells end up in either of these two positions.

At an early stage, cells in the inner cell mass are not determined; they are not yet specified to form brains or toes. If a single marked cell from the inner cell mass of one mouse embryo is injected into the inner cell mass of another it will become integrated into the development of the embryo and respond to developmental signals. If it carries a marker its descendants can be recognized, and they can be seen to give rise to all sorts of tissues, from muscle and bone, to liver and brain. This is consistent with the early embryo showing remarkable capacity to regulate when perturbed. Even when the embryo begins to gastrulate, if up to 80 per cent of the cells are killed off with a drug the embryo can regulate and quite normal mice can develop.

A GOOD LINEAGE

Nature seems quite profligate in the number of ways it has arranged for embryos to construct organisms. We may find unifying principles in patterns of gene activity, gradients, signals, modes of movement, but there is still much variety for which we have no explanation whatsoever.

Consider, for example, the early cleavage of the embryo. This is a relatively simple process and divides the egg up into a number of smaller cells from which the embryo will develop. It is a puzzle why there should be two main classes of cleavage patterns—radial and spiral. Radial cleavage is clear and straightforward and occurs in sea-urchins, amphibians, and some invertebrates. The cleavage planes are at right angles to one another. Many invertebrates, on the other hand, such as snails and worms and crustacea, have a spiral pattern of cleavage. In spirally cleaving eggs the cleavage planes are oblique so that the cells, when viewed from above, take on a spiral arrangement.

In addition, the pattern of cleavage is both highly ordered and complex and there are fewer cells at the time when gastrulation begins. Because there are so few cells and a well-defined pattern, it is possible to follow the lineage and fate of every cell at least up to the time of gastrulation. The patterns of spiral cleavage are hard to understand in functional terms, but one example does, however, have a tantalizing correlation with adult forms. In molluscs, the direction of coiling of the shell corresponds with the spiral arrangement of the cleavage. Snails with right-handed spiral cleavage have shells which coil with the same handedness, and left-handed shells have a left-handed spiral cleavage.

Spiral cleavage focuses our attention on the role of cell lineage in specifying pattern. Lineage itself could specify cell fate. That is, when a cell divides, the daughter cells may be dissimilar and already set along different developmental pathways, and so no cell-to-cell inter-actions need be involved. If the embryos have taken so much trouble to have so well-defined patterns of cleavage it is a reasonable expecta-tion that the cleavage pattern may be involved in specifying cell fate. In general terms that is true. Spirally cleaving embryos show much less regulative ability than that seen in, for example, sea-urchins and amphibians. There seems to be much less interaction between the cells. Fate seems rather to depend more on ancestry than position, family rather than friends. We are almost back again with the problem of preformation, but in a modern form. The best understood spirally cleaving animal is the nematode, *Caenorhabditis elegans*.

THE WORM'S STORY

The intensive study of the nematode *Caenorhabditis elegans* derives from the molecular biologist Sydney Brenner. In 1974, he decided that none of the developmental systems being studied really gave him what he needed for a study of development. He picked out the small nematode that had just the right characteristics, and single-handed, he established a thriving field from scratch. The adult worm is about 1 millimetre long and consists of only about 1000 body cells and thousands of germ cells. In fact, the male has exactly 959 body cells and the female 1031. Like higher animals it has a nerve cord, muscles, and a gut, with a brain at the mouth (hardly a head) end. It only has 3000 genes and by this criterion is about 20 times more complicated than a bacterium and 40 times less complicated than a human. It is excellent for 'doing genetics'.

The complete lineage of every one of the cells has now been followed. The pattern of cell divisions was found to be invariant, every

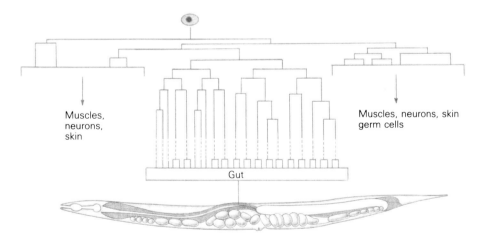

normal worm going through exactly the same pattern of cleavages to form the body. But there is no clear logic to the complex pattern of cell divisions which divide up the egg and generate the worm. There is, thus far, no global plan apparent which we would regard as logical and satisfying. At most we can make some sense of small parts of the pattern.

Consider the different cell types, such as muscle, skin, gut, and nerve. Do each of these cell types derive from just one stem cell? That is, do all the muscle cells descend from just one or two ancestral muscle cells? The answer is no. These cell types can derive from several cells well separated in the lineage. In fact, occasionally, a nerve cell and a muscle cell can be sisters derived from a common mother cell.

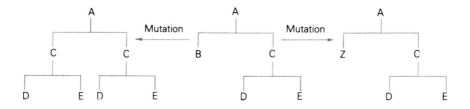

One way to think of the development of the worm is to draw an analogy with a digital computer and think of the cells at every cell division making a discrete choice as to what to do next. One must also accept that the programmer of the cells had a different logic from ours, or at least a curious sense of humour. Some of the mutations that have been found support this digital computer image by altering the program in different ways. For example, a typical lineage may result in cell type A giving two cell types B and C, and only C continues to divide to give types D and E. Mutants can result in the substitution of type Z for B; or make the lineage symmetrical—B now giving D and E; or even generating a stem cell line with B being replaced by A which repeats the pattern of cell divisions and cell differentiation.

The emphasis on cell lineage in the worm is a little misleading since the implication is that the behaviour of the cells is entirely determined only by the lineage and not by cell signals. In some cases this is true, but in other cases it is not. A technique used to test for the autonomy of cell development is to see what happens to cells if their neighbours are killed-off by a fine laser beam. If development of the cell really is autonomous then killing-off neighbours will have no effect, however if it is dependent on signals from adjacent cells its development will be perturbed. When the laser is used to kill a cell the adjacent cells often continue to develop normally showing that signals are not necessary. But in some cases there is dependence, and laser killing shows that, for example, one kind of cell induces three other cells just next to it to develop along a particular pathway. Interactions also specify a special fate for the central cell.

In spite of all the intensive study of the worm's development there is still little, if any, understanding of why two sister cells can develop, autonomously, along different pathways. In the absence of environmental signals there has to be the unequal distribution of some factor at cell division so that the cells behave differently from one another. One possibility is that there are special cytoplasmic factors that are asymmetrically distributed at cell division—at very early stages there is very good evidence that factors in the cytoplasm are unequally distributed during cleavage.

Not quite a homunculus

If the cytoplasm of the egg has a well-defined pattern of constituents which determines the later organization of the embryo then cleavage may just be a way of partitioning this cytoplasmic pattern in the egg so that specific constituents go to specific cells. These constituents would then specify how the cells develop. In this way, the basic pattern in the egg could be transferred to the cells. There would be little, if any, need for cell conversations, gradients, and other organizing mechanisms. The egg would almost be a homunculus. The cells become the cytoplasm they inherit.

The development of the tunicate (sea-squirt) egg, which has a radial pattern of cleavage, illustrates this cytoplasmic prepatterning. Certain species have, very fortunately, coloured cytoplasm which is found in specific locations in the egg. Each coloured region ends up, during cleavage, in specific cells and those which acquire similarly coloured cytoplasm differentiate in the same way. Muscle cells come from cells that contain the yellow cytoplasm that is present as a crescent-shaped patch in the egg, a grey patch in the egg cytoplasm gives rise to the nervous system, and so on. The dependence of muscle development on specific cytoplasmic factors is dramatic. At the eight-celled stage the yellow cytoplasm is confined to a pair of adjacent cells. When the

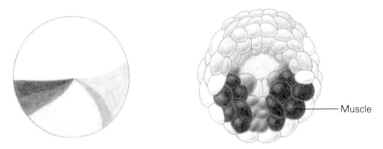

embryo is divided up into pairs of cells then only the pair with yellow cytoplasm gives muscle. But if, by manipulation, some of the yellow cytoplasm is squeezed into adjacent cells they too will develop muscle, something they would never do normally.

Lineage-based development, with its dependence on cytoplasmic localization, can account for much of the sea-squirt development but the idea should not be taken too far. There are some cell interactions, and in some species, cells lacking the yellow cytoplasm can form muscle. Even so, it is a very impressive example of the role of cyto-plasmic factors and cell lineage.

SELF ORGANIZATION

In marked contrast to lineage mechanisms are those mechanisms which might generate pattern spontaneously.

The English mathematician, Alan Turing, was an extraordinary polymath. Not only did he lay the foundation for the theory of computer programs, but in 1952 he published a highly original paper on pattern formation. He showed that it was possible, starting with a uniform concentration of chemicals which interacted with each other, for the system to develop differences in the concentration of the chemicals, which he called morphogens, such that there would be chemical waves with peaks and troughs of concentration of the morphogens. The system was self-organizing and the pattern would occur spontaneously. Turing suggested that the concentration of the chemicals might control the pattern of a developing system. If five peaks were set up they could provide the basis for the tentacles of hydra, or for the fingers in the early development of the human hand.

Turing's model was mathematical. Since that time it has been explored in chemical systems with further mathematics and also experimentally. In essence, the systems involve diffusing chemicals that interact with one another—hence their name—reaction-diffusion models. For example, the system may use a rapidly diffusing inhibitor and a slowly diffusing activator; the inhibitor inhibits activator production but its own production is dependent on the activator itself;

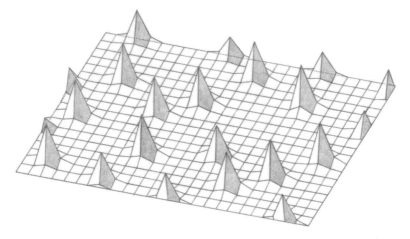

and the activator stimulates its own synthesis. For particular conditions wave-like patterns will be generated. There are chemical systems which mimic it in some ways. In a complex mixture, whose chemistry is well understood, the Belousov–Zhabotinsky reaction can occur. When the reaction is carried out in a dish, complex moving patterns of coloured bands spontaneously arise, including concentric rings and spirals. On occasion, a pattern of spots appears.

Reaction-diffusion mechanisms are very good at generating repeated patterns of many peaks of morphogen concentration. Some of these patterns, as the Oxford mathematician, Jim Murray, has shown, resemble the coat patterns of animals like leopards and zebras, and the spots on butterfly wings. Reaction-diffusion could well be the way they arise in development. Such mechanisms are very sensitive to the size and shape of the system in which the reactions take place because the boundaries are very important. In a system which gives spots, making a region long and thin results in stripes—striped tails are a common feature of many animals.

The great advantage in generating patterns in development by such reaction-diffusion mechanisms is that they are self-organizing and so can generate patterns without the complex programming that the interpretation of positional information requires. They can also provide the spontaneous formation of gradients. Unfortunately, attractive as such mechanisms are, direct evidence for their involvement is very weak.

FEATHER PATTERNS

There are few patterns as varied and beautiful as those made by the feathers of birds. The patterns largely reflect the distribution of pigment-containing cells, but some of the colours, like the blues of the peacock's tail, come from the way the structure of the feathers refracts light. Patterns of pigmentation are due to the distribution of cells that are derived from the neural crest (Chapter 2). Those cells which can form pigment migrate beneath the skin and enter all the feather germs. These germs are small outgrowths on the skin arranged locally in a hexagonal pattern. Whether or not the cells that enter the feather germs make pigment is controlled by the feather germs, each of which has its set of positional values—each feather has its own address. Depending on their positional value the feather germs promote pigment formation by the migrant cells, or suppress it, and in this way the global pattern is specified. The positional values that are used for pigment development in the wing are the same as those used for cartilage, muscle, and tendons (Chapter 4). In this way, the various elements in the wing's positional field, for example, are correctly placed in relation to one another. More generally, the positional field in the skin enables a wide variety of pigment patterns to be generated—it is almost like painting by numbers.

Within each feather there is a further patterning which determines the distribution of the pigment. Pigment granules are extruded from the pigment cells as the feather grows out, and changes in the timing of extrusion leads to barred feathers. More complex processes, which are not understood, produce local spots in the feather, as in, for example, some canaries.

Epigenesis

Aristotle was right for the wrong reasons. He coined the term 'epigenesis' to describe the emergence of new structures and forms during development, but his justification for this view was based on philosophy not experiment. It should now be very clear that development is a dynamic process in which cells repeatedly interact with each other, change shape and position, and become different. Each stage of development effectively sets the stage for the next. Cell movements, for example, may bring tissues in apposition resulting in new interactions leading to further movements. Other 'flags' can be thought of as developing within one part of the French flag. New positional fields are established after the main axis is set up—those that give rise to the limbs will be examined in the next chapter. In a way, development can be thought of as a cascade, one event leading to another. Sometimes the sequence is confined to the changes in single cells, at other times groups of cells have multiple influences on their neighbours. This, in turn, influences yet other cells.

By concentrating on spatial aspects of the developing system the temporal aspect has been neglected. The changes with time are just as important as those that occur during spatial patterning and are indeed part of the process. Epigenesis is indeed a process set in time. Cells, for example, are only competent to induce, or be induced, for limited periods. Unfortunately, we understand less about temporal than spatial processes in development.

FINGERS AND TOES

THE DEVELOPMENT of arms and legs is very important in its own right, and it provides one of the best model systems for analysing how an organ develops once the body plan has been laid down. Not only is the pattern of cartilage elements in the early limb, such as the humerus, radius and ulna, and digits, quite simple, but the embryonic chick limb bud is quite easily accessible to experimental manipulation and is an excellent model for the human. The basic pattern consists of one main element, the humerus, followed by two smaller elements, the radius and ulna, then some small structures in the wrist, and finally the digits. Overall, it has a pleasing simplicity and there is a similar pattern in the leg. By contrast, for example, the development of the heart is much more convoluted, involving as it does the complex folding and fusing of tubes. Of course, in addition to the bony structures in the limb—which are first laid down as cartilage and only later replaced by bone—there are muscles and tendons whose pattern is more complex.

The wings and legs develop late. The main body axis is well advanced, and is already segmented into somites; the head too is moderately advanced and the eyes are prominent. The forelimb begins to develop first; a small bulge appearing on the flank of the embryo—the first signs of the limb bud. It is made up of an outer sheet of cells which forms a particularly thick structure—the apical ridge—at the tip, with an inner mass of loosely packed and rather dull looking cells. The cells multiply and the bud grows out in a paddle-like form. About half a millimetre behind the tip, the cells in the core become denser and begin to make the cartilage of the first element in the limb, the humerus. With further outgrowth the radius and ulna develop within the inner mass of cells, then the wrist, and finally the hand with its digits. After 10 days of development within the egg, the forelimb of the chick—the wing—has a

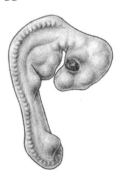

pattern very similar to that of the human arm, except that there are only three fingers. It is very convenient that the three digits are so unlike one another since they can be easily recognized. Based on evolutionary considerations, the small anterior digit is designated digit 2, the large central one as digit 3, and the posterior, middle-sized digit, as digit 4. The assumption is that, in evolution, digits 1 and 5 were lost from the primitive five-fingered limb.

Our model for the patterning of the elements of the limb is based on the idea of positional information, that is, the cells in the limb have their position specified and this determines how they will behave. The positional field in the limb determines, coherently, the pattern of all the structures—cartilage, muscle, tendons, feathers, and nerves. Although the limb is three-dimensional I will treat it as if it were two-dimensional and only consider position with respect to two axes, the antero-posterior axis which runs across the limb from digit 2 to digit 4, and the long, or proximo-distal axis, which runs from the shoulder to the tip of the digits. How do the cells acquire positional information?

Outgrowth of the chick limb bud is mainly due to cell multiplication at the tip of the limb just beneath the thickened ridge-like structure in the surface layer. We called this region of proliferation the progress zone because it is there, we believe, that the cells acquire their positional values. The special nature of the progress zone is due to a signal from the overlying ridge. The cells in the progress zone all multiply and can respond to positional signals. Only when the cells leave the zone do some begin to differentiate into cartilage; and, as just stated, the cartilage elements are laid down in a proximo-distal sequence—first humerus, then radius and ulna, and only then wrist, and finally hand. In our model, the cells in the progress zone have their position specified along the two axes by two different mechanisms. For the proximo-distal axis, our suggestion is that the cells learn their position by measuring how long they remain in the progress zone, a mechanism based on measuring time. For the antero-posterior axis there is evidence for a positional signal from the posterior margin of the bud. I will first consider the signal along the antero-posterior axis and then the 'timing' mechanism for the other axis.

POSITIONAL SIGNALLING

At the posterior margin of the chick limb bud there is a small group of cells known as the polarizing region. They look no different from any other cells in the limb bud but they have the ability to specify the position of the cells along the antero-posterior axis. The polarizing region thus has properties similar to that of the amphibian organizer. A key experiment shows this. A hole is made in the shell and a small cube of cells is carefully cut out from the posterior margin containing the polarizing region and grafted into the anterior margin of the limb bud of another embryo. After this the hole in the shell is sealed and the embryo is allowed to develop for a few days. The host limb now has two polarizing regions and develops a second set of digits which are a mirror image of the normal ones. Instead of the normal pattern of digits, 2, 3, 4, the pattern is now 4, 3, 2, 2, 3, 4. Grafting an additional

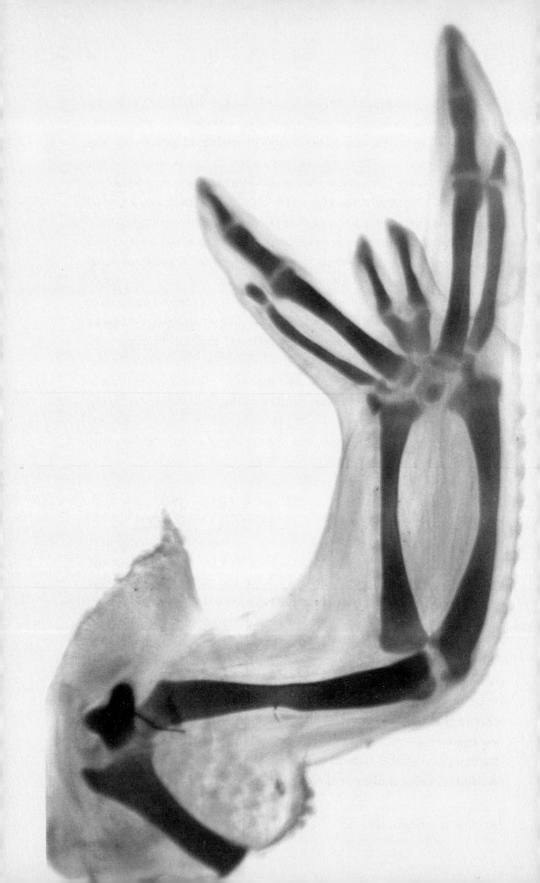

polarizing region into the anterior margin has done two things—the limb bud has widened and the polarizing region has re-specified the position of the cells in the anterior part of the limb so that they form additional digits. These extra digits do not come from the graft, but from the cells adjacent to it in the anterior part of the bud. Our model proposes that the polarizing region normally provides a positional signal along the antero-posterior axis. This signal, we believe, is a

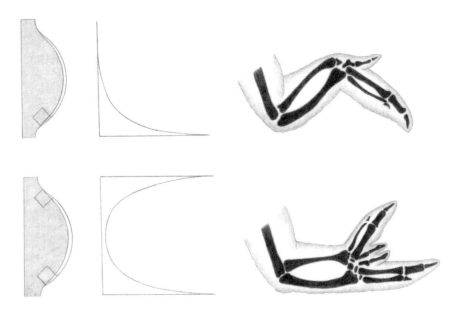

diffusing chemical, a morphogen, which is released by the polarizing region, so the concentration is highest at the posterior margin and the concentration decreases away from the polarizing region, being lowest at the anterior margin. This morphogen concentration gradient can thus be used to determine the position of the cells. If digit 4 develops at a high concentration, digit 3 at an intermediate concentration, and digit 2 at the lowest concentration, then one can visualize how both the normal and mirror-image patterns arise. When we add another polarizing region to the anterior margin there are now two sources of the chemical and a U-shaped gradient in the chemical is established, resulting in the embryo's limb having a pattern of digits 4, 3, 2, 2, 3, 4.

Mirror-image embryonic chick limb after graft of a polarizing region

We devised a test for this model—and it is just a model—by trying to weaken the signal from the grafted polarizing region. If, we reasoned, there really is a morphogen gradient, it should be possible to weaken the signal by decreasing the concentration, and so, instead of getting an additional set of digits 4, 3, 2 at the anterior margin only a 3, 2 or a 2 should develop. This is exactly what we found. We weakened the signal in several ways, one of which was by damaging the cells of the polarizing region with high doses of X-rays before grafting it to the anterior margin. As the dose of X-rays was increased, so the pattern of additional digits changed from 4, 3, 2 to 3, 2 to 2. Another approach was to graft fewer and fewer polarizing region cells and again the result was the same.

An obvious question is the nature of the morphogen. The idea that graded morphogen concentrations are important in controlling development is a very old one and the attempts to identify the morphogen have failed, almost without exception. It is a technically very difficult problem compounded by the fact that one has no idea of what to look for, and the quantities may be minute. First, we tried a direct approach and ground up polarizing region cells and put the mixture into agar blocks which we then grafted to the anterior margin. The hope was that the morphogen in the mixture would slowly leak out and specify new digits. If that happened we could then try to purify and isolate the morphogen from the mixture—but we had no luck. Therefore, we adopted a raffle-ticket approach and tried a variety of pure chemicals, just guessing which to choose, on the grounds that if one holds a raffle ticket one's chance of winning may be small but at least it is finite, and greater than zero. If one does not have a raffle ticket one's chances are zero. Therefore, everyone in the laboratory was invited to test their favoured compound. Using this approach we believe we may have discovered the signal.

It would be nice to pretend that our decision to try retinoic acid, a derivative of vitamin A, was based on sound chemical principles. This was not the case. We chose retinoic acid because I met a friend, at a meeting, and he told me that it affected cell communication. In addition, retinoic acid is insoluble in water and so would remain where we put it in the limb for some time; this was important because we already knew that to exert its effect the grafted polarizing region needed more than 12 hours. To our surprise and delight it mimicked the effect of the signalling region. If a special tiny bead was soaked in retinoic acid and then grafted to the anterior margin, a mirror-image limb developed with a pattern of digits 4, 3, 2, 2, 3, 4. If the concentration of the retinoic acid was reduced, then, just as happened when we weakened the signal, from the polarizing region, the pattern was 3, 2, 2, 3, 4, then 2, 2, 3, 4, ending up with 2, 3, 4. Unfortunately, this does not prove that retinoic acid is the signal: it could be that it merely mimics the true signal. But other workers have now shown that retinoic acid is actually present in the limb, and its concentration is graded in just the right direction. In addition, special receptors for retinoic acid have now been identified in the cells in the progress zone. All of which is in favour of retinoic acid being the signal, but the evidence is not unequivocal.

Whatever the ultimate nature of the signal turns out to be, there is another question to be asked. To what extent is the signal in the chick the same as in other vertebrates? How conservative or parsimonious has nature been with its signals and its building materials? If a piece of tissue is taken from the posterior margin of a mouse limb bud and grafted into the anterior region of a chick limb bud, it too can specify additional digits, but the digits are, of course, chick digits. A similar result has been obtained by other workers who grafted the polarizing region from an aborted human embryo into the chick. It is clear that this signal is the same in all higher vertebrates; what has changed in evolution is the response. It is that response that makes the digits of birds, mice, and men different.

POSITION AND TIME

For the proximo-distal axis, that is the long axis of the limb, from the shoulder to the fingers, there is a rather different mechanism for specifying positional information. It is a mechanism based on time. The cells in the progress zone seem to be able to measure how long they remain in the progress zone and this enables them to 'know' their position along the proximo-distal axis.

All the cells in the progress zone at the tip are multiplying and, therefore, cells are continually leaving the progress zone as the bud grows out and there is a trail of cells left behind. The first cells to leave will form proximal structures, like the humerus, whereas those that remain in the progress zone longest will form the tips of the digits. If the cells have a 'clock' that stops when they leave the zone the time on the clock will give them their position along the axis. The longer the clock runs the more distal the position of the cells.

This mechanism accounts at once for the observation that limbs are truncated when the thickened ridge at the tip of the bud is removed. Since the progress zone's existence is dependent on a signal from the ridge, removal of the ridge results in the disappearance of the progress zone and so effectively the clock in the cells is stopped permanently.

No further distal positional values can be generated. Thus, if the ridge is removed from a very early bud only the humerus will develop; but if it is removed at a later stage the radius and ulna and wrist will develop but the digits will be absent. Whenever a limb abnormality involves truncation, one can be pretty sure that this is due to damage to the ridge and so loss of the progress zone.

We do not have good evidence to support the timing mechanism. However, the model is consistent with experiments in which we blocked proliferation in the progress zone or killed off some of the cells. When this happens, cells stay much longer in the progress zone than they normally would, and very few cells leave the progress zone at early stages. With time, normal proliferation is restored in the pro-gress zone with the result that proximal structures are absent and distal structures are quite normal. If cells in the progress zone of the mouse limb bud or the chick limb bud are damaged often merely a hand develops attached to the shoulder and this may provide a basis for understanding the effects of thalidomide.

Thalidomide, as many will recall, is a drug that was put on the market in 1958 to calm the anxious and to overcome morning sickness in pregnant women. It was marketed under the name Distaval and the advertisements in 1961 indicated that it could be given with complete safety to pregnant women. An Australian paediatrician, William MacBride, was the first to warn the medical profession of the danger of thalidomide. He had come across severe abnormalities—particularly of the limbs—in children of mothers who had taken thalidomide during pregnancy. The drug was eventually withdrawn but there are about 300 young adults alive today who were affected. Three-quarters of these have limb defects—their arms and legs are shorter, and sometimes missing altogether. In some cases, there is a small hand attached to the shoulder.

No one really knows how thalidomide causes such abnormalities. The doses given to pregnant women which caused the effect were extremely small. When researchers tried to mimic the results on other animals, such as chick embryos, it did not have the same effect. However, it does cause similar abnormalities in monkeys and there is evidence that it may cause blood vessels in the limbs to leak, which leads to severe damage of the surrounding cells. As we have seen, damaging cells in the progress zone of the early bud causes proximal structures to be lost, and so our model just might provide a clue as to how thalidomide works. The hand attached to the shoulder in some thalidomide children is just like the isolated 'hand' that develops from chick embryos whose cells in the progress zone have been damaged.

ARMS AND LEGS

The arm and the leg of many animals are both very similar and very different. They have a very similar pattern of elements—the humerus is like the femur the fingers are similar to the toes—but the detailed form is clearly very different. So, it is not too surprising that the signals and mechanism of development are virtually identical in the arm and the leg. For example, the polarizing region from the leg will produce extra digits, and a wing-polarizing region in the leg will specify additional toes. The difference between the arm and leg is thus due to the different interpretation by the cells of the same positional signals. An elegant illustration of the similarity in signals, but difference in response, is provided by grafting tissue from the proximal part of an early leg bud that would normally develop as the femur, into the progress zone of an early wing bud. The grafted tissue now develops into toes. The grafted cells which had a proximal character are now in the progress zone again and so acquire more distal positional values. Because they retain their leg-like character and since their positional value is changed from proximal to distal, toes form.

The different interpretation of positional signals by arms and legs can be traced back to their different developmental history, each bud arising at a different level along the main body axis. Their position along this axis alters the cell's internal programme such that they now interpret the positional signals within the bud so as to make the arm and leg different. It is assumed that different pattern-controlling genes are switched on in forelimb and hindlimb, but this remains speculation. However, in insect development there is excellent evidence for such a mechanism (Chapter 7).

A PREPATTERN AND THE SIXTH FINGER

The limbs of vertebrates can be very different—compare that of a bat with its long fingers supporting the web for flying, with the long pole-like structure of a horse's leg, where all the digits except one seem to have disappeared. However, many vertebrate forelimbs have a similar basic structure—a single element, the humerus, followed by two elements, the radius and the ulna, and then three or more elements in the wrist and digits. Given that consistent pattern, it is hard to resist the idea that there is some basic mechanism that lays down first one element, then two, and then three, and so on. This is a temptation one should in general resist, as I regard it as a cardinal rule in developmental biology not to try to infer developmental mechanisms from final forms. Perhaps in this case the injunction can be ignored since the pattern is so widespread and so instinctively attractive. Moreover, there are no counter-examples; there are no limbs which start with two elements followed by a single one. More importantly, there is experimental evidence for a mechanism that generates this basic pattern and which does not require anterior-posterior positional information.

The key experiment is to take early limb buds, remove the covering jacket, separate the cells, mix them up, and then pack them back into the jacket which is then grafted to the flank of an embryo. Such buds containing re-aggregated cells do not develop normally but they can form jointed cartilage elements and sometimes very good-looking digit-like structures. Since there is no signalling region to specify position along the antero-posterior axis—the cells being all mixed up, they are clearly capable of self-organization. They do, of course, have a progress zone, because the jacket has a thickened ridge at the tip. Even so, the capacity for self-organizing rod-like cartilage elements is impressive. A possible mechanism whereby the disaggregated and

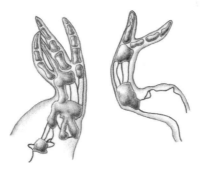

reconstituted bud could form limb-like structures might involve a reaction-diffusion mechanism of the type invented by Turing, as discussed in the previous chapter. Such a mechanism could play a role in normal limb development, laying down a basic prepattern of cartilaginous elements: first one element, then two, then three, and so on. Reaction diffusion could provide a series of chemical waves generating first a wave with a single peak, followed by one with two peaks, and then one with three peaks. Then, if cartilage is made only where there are peaks, a basic pattern will be established which could then be modified by positional information to give the characteristic patterns of the humerus, radius and ulna, wrist, and hand.

Such a model would be able to explain the not uncommon occurrence of people with six rather than five fingers. There would normally be five peaks where the fingers were forming but if the limb bud, accidentally, was wider than normal, a sixth peak could form, giving the sixth finger. It is a characteristic of the wave-generating mechanism that the number of peaks is dependent on the size of the system and a moderate widening could allow another peak to form. The possibility of a self-organizing patterning mechanism is attractive but evidence for it is, at present, minimal.

MUSCLES AND TENDONS

The cartilaginous elements which will eventually become the bones of the limb are just one of its structural elements. Muscles and tendons are just as important and form a much more complex pattern. There are over 30 muscles in the chick limb which attach to the cartilage at

one end and to tendons at the other. The tendons, in turn, are like cords that transmit the force of the muscle to other parts of the limb skeleton. They too attach to very specific parts of the limb. The pattern of both the muscles and the tendons of the chick limb respond to the same positional signals as the cartilage. For, when a polarizing region is grafted on to the anterior margin of a bud it is not only the cartilaginous elements that form a mirror-image duplicate but the muscles and tendons are also duplicated. The same positional information is used to specify all the structures in the limb.

It is typical of vertebrates that the muscles which move the fingers are quite remote from the fingers themselves. If you look at a dissection of the human arm you can see that the string-like tendons which attach to the fingers are joined further back in the arm, to the muscles. It is almost as if the muscles controlled the fingers as marionettes are controlled by their strings. Although there is an essential functional relation between skeletal element, tendon, and muscle, it does not follow that their initial development is similarly linked.

Each of the elements of the chick limb shows a striking early sense of autonomy. Each seems to be specified separately and to initially develop quite independently of its future neighbours and associates. Consider the long tendon that attaches at one end to the muscle *flexor digitorum profundus* and at its other end to digit 3. As its name implies the muscle, when it contracts, will flex the digit. To what extent is the development of the tendon dependent on the muscle? We tested this by grafting the tip of a limb bud to the flank of the embryo. There the tip developed as a hand but all the more proximal structures were absent, and so there were no muscles proximal to the hand. Nevertheless, the tendon developed in the right place even though it had no muscle to which it could attach. This is a typical result and it seems that, like the tendons, the early development of the muscles is quite autonomous. The early development of both are controlled by their position within the limb. But, in addition, the muscles and tendons have a mechanism whereby they join up with whatever element is next to them. In this they are quite promiscuous and unselective and tendons will join up with whatever muscle they find at their end.

The qualification of 'early' in relation to autonomy is very important, for the later development of muscles and tendons is not autonomous. Although the long tendon attaching to digit 3 will develop initially, even though it does not attach to a muscle, it will not persist unless it does attach. If it is not placed under tension and pulled it will disappear. Similarly, the muscles will not grow in length unless they are attached to tendons and bones so that as the bones lengthen, they are stretched. The later growth of the limb's elements is co-ordinated by mechanical interactions with the growth of the bones providing the driving force. This ensures that muscles are the correct length for the bones to which they attach (Chapter 10).

Patterning in all systems occurs in small groups of cells, the maximum dimension of any system in which patterning is taking place does not exceed a millimetre. The limb is no exception, which means that the basic pattern of the limb is laid down on a very small scale. Growth of this basic pattern is required to generate the mature limb. The growth characteristics of the bones are laid down at the time when the pattern is specified and the elements are very small and this early specification will control growth for many years (Chapter 10).

CELL DEATH

Cell death is a normal feature of limb development and helps sculpt the form of the limb. For example, in mice and humans the digits are initially joined together and only cell death between the digits separates them. It is not that the cells die because they are sick or somehow abnormal—dying is part of their programmed development.

It is, in principle, no different from cartilage differentiation. Cell death also occurs in the chick at the anterior and posterior margins of a late bud, helping to mould its overall form.

In 1968, the American embryologist, John Saunders, was interested in finding out whether this cell death was programmed early on, or whether it was dependent on interactions with the surrounding cells. So, he grafted tissue from the posterior margin that would die to other parts of the chick limb to see if he could rescue the cells from death. The results were equivocal, but to his surprise he found that when he grafted the tissue to the anterior margin of the limb there was a dramatic change in the development of the limb—a whole new set of digits developed: he had discovered the polarizing region. The signalling properties of the polarizing region are now known to be quite unrelated to cell death which is a separate process.

VARIATION ON A THEME

The limbs of different vertebrates such as human, dog, bat, and horse, are variations on a basic theme of an increasing number of elements as one proceeds from shoulder to hand, from the single humerus to the five fingers of the hand. This basic pattern has been modified in evolu-

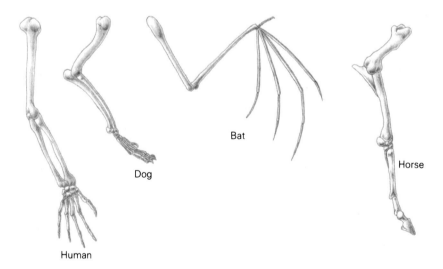

Bat

Dog

Horse

Human

tion to give a limb that serves quite different functions. In the bat, the five fingers have been programmed to be long in order to support the web for flying. In the horse, the digits have been progressively reduced so that just the central core carries the load and the two adjacent ones are present as splints—remnants of the digits that once were there. In cattle, two digits have fused for load carrying. In whales, the digits are longer in the hand and the hindlimb has disappeared. It is only because the different elements in the limb have their own positional identity that their pattern and form can be altered independently during evolution. How the genes control these changes is a central problem.

The changes are not due to alterations in the signals during limb development but due to the alterations in cell response. There has been a change in the genes, which alters the genetic programme. For example, in certain limbless lizards and snakes a limb bud develops but the apical ridge dies thus stopping limb development.

THE CELL'S PROGRAMME

The models proposed for the development of the limb place quite a big burden on the cells' response to apparently simple signals. Cells in the limb bud must record whether they are forelimb or hindlimb cells and so alter their programme. They must also record their position in the limb, and this may mean that they have both to measure accurately the concentration of a morphogen, and measure time. Again, they have to interpret their position according to their genetic programme and all these processes are controlled by genes, which still have to be identified. That such genes exist is shown by the many mutants in the genes of mice that can result in abnormal limb development. These abnormalities range from the complete absence of limbs to loss or gain of digits.

Very recent work has provided, for the first time, an indication of how the position of cells in the developing limb might be recorded. There is a set of genes involved in patterning the embryo, the homeobox genes (Chapter 7), which are expressed in the developing limb buds in just the way that would suggest that they may encode positional information. For example, successive homeobox genes can be seen to be active, the further away from the polarizing region the cell is. This observation may at least provide a clue to the molecular basis of positional information.

By understanding what the cells do during limb development, we can begin to ask how the genes control these activities. Thus it is necessary to examine the cell more closely and to try and understand its internal programme and how it responds to external signals. It will become evident that the cell is, in a way, more complex than the embryo.

EX DNA OMNIA

E X OVO OMNIA was on the frontispiece of William Harvey's book on embryology in 1651. Ex DNA omnia is more appropriate since development is dependent on DNA. The genes are made of DNA, which is both simple and complex, passive yet active, a truly magical molecule. It is the DNA, and the DNA alone, that carries the genetic information. But DNA is a rather passive molecule even though it rules our development. In contrast, it is the very active proteins that are the true wizards of the cell. The power of DNA lies in its containing both the instructions for making all the proteins in the cell, and the programme which controls their synthesis.

DNA is a very long string-like molecule and it is packaged, with special proteins, in the form of chromosomes within the nucleus of the cell. Because it is such a long molecule the DNA is folded and twisted within the chromosome. Even so, chromosomes are themselves long and thin and are not normally visible in the light microscope. However, when the cell divides the chromosomes condense—this makes them easier to distribute to the daughter cells—and are now easy to see. Humans have 46 chromosomes, 23 from the father and 23 from the mother, each of which can be matched with its partner from the other parent. Each chromosome contains just one DNA molecule

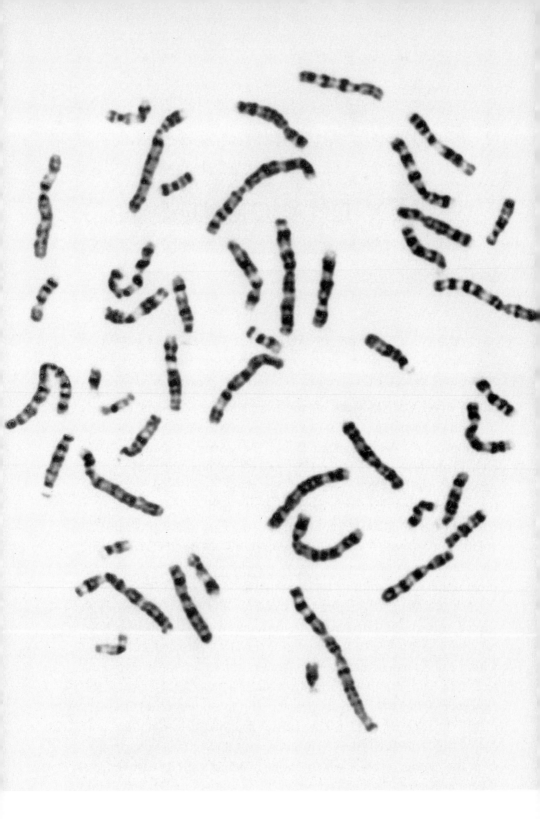

Human chromosomes

so there are exactly 46 molecules of DNA in the fertilized human egg and all normal body cells.

The DNA molecule is made up of a sequence of four basic units—nucleotides—strung together. There are two strands of nucleotides in every DNA molecule that coil around each other to give a double helix, the nucleotides in one strand pairing with those in the sister strand. The order of the nucleotides along the molecule provides the information for making proteins—and a stretch of DNA that specifies a single type of protein is a gene.

PROTEIN VERSATILITY AND SEQUENCE

The nature of a cell is governed by the proteins it contains. Proteins are capable of performing the most amazingly diverse functions within the cell. All the thousands of essential chemical reactions in the cell—the conversion of energy, the building of new molecules, the breakdown of molecules—depend on enzymes which enable these reactions to occur. All enzymes are proteins. Proteins also provide the structural basis for the movement of muscle, the transmission of messages by nerves, and give strength to skin and tendons. The red blood cell can carry oxygen because it contains the protein haemoglobin, muscles can contract because of the proteins, actin and myosin. Skin cells owe their toughness to keratin, tendons are made of the protein collagen.

There are many different cell types in the human body: liver cells, red blood cells, nerve cells, skin cells, and so on. Some are obviously very different but others look rather similar. If you wish to identify a cell, the key question to ask is 'What sort of luxury proteins do you have?' If the reply was 'haemoglobin' you would be dealing with a red blood cell; albumin would indicate a liver cell; insulin a cell from the pancreas. The reason you would ask about 'luxury' proteins is that these proteins are special to the cell. 'Housekeeping' proteins are common to most cells and are needed to carry out all those basic functions that most cells require, such as the production of energy. Housekeeping proteins are just as important to the cells as the luxury ones, but they are not the markers of cellular individuality and for development are less interesting..

Proteins are essentially long strings of units called amino acids. There are 20 kinds of amino acid and the order of the amino acids, which is always the same for a given protein, determines the properties of the protein. The order of the amino acids determines how the protein 'string' will fold up. Some proteins, like collagen in tendons, are long and thin and have rope-like properties; others which act as enzymes fold into complex shapes which enable them to bind to other molecules.

DNA CODES FOR PROTEINS

DNA codes for the sequence of amino acids in proteins. The sequence of four nucleotides along the DNA molecule constitutes a gene that codes for the sequence of the 20 different amino acids in proteins. This code is in the form of triplets, a different combination of three nucleotides corresponding to each amino acid. So, knowing the sequence of nucleotides in a gene, which may be thousands of nucleotides long, the order of amino acids in the protein can be worked out.

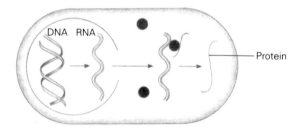

DNA is confined to the nucleus, but protein synthesis takes place in the cytoplasm. The code for a particular protein is first copied by a process known as transcription into a complementary molecule, messenger-RNA, which is also made up of four nucleotides. This messenger-RNA then carries the code into the cytoplasm where the protein is synthesized.

One can think of the DNA in the nucleus as an enormous reference library containing the instructions for making all the proteins. There are about 50 000 different proteins in our bodies. The DNA contains not only the instructions for making all the proteins but is also involved in the controlling of which protein should be made when and where. Contained in the DNA are 'library rules' dictating what has to be done for the instructions for a particular protein to be taken out and 'read'. These rules are the core of the genetic programme for development.

MUTATIONS

A change in the sequence of the nucleotides in the DNA of a gene is a mutation. This can be the result of a loss of part of the DNA or the replacement of one or more of the four nucleotides by others. Another cause is that complete regions of DNA are transposed from one part of the molecule to another, or that foreign DNA from a virus has been inserted. In each case, the normal sequence of nucleotides is altered, and the alteration can result in abnormal development. Abnormal development may occur if a protein is not made at the right time or place, or the protein is defective or if too little or too much is made. A clear example of a mutation altering development is the inherited genetic defect, sickle cell anaemia.

In people with sickle cell anaemia, the molecule that carries oxygen in the blood, haemoglobin, is abnormal, and the red blood cells take on a sickle shape. The origin of the defect lies in the gene coding for haemoglobin which is 1600 nucleotides long. A change in a single nucleotide, the 67th along the gene, results in a single incorrect amino acid being inserted into the protein chain. This in turn causes the protein to fold up in an incorrect way, which results in the haemoglobin molecules sticking to each other. They now spontaneously assemble into rods which press against the membrane of the red blood cell deforming it from a rounded into a sickle shape. Because the cells are sickle cell shaped and rather rigid they have difficulty passing through fine blood capillaries, and this results in the tissues not having enough oxygen, causing anaemia. So the change in just one nucleotide in the DNA can lead to change in shape of a cell with resulting anaemia. The route from gene to observed effect can be very tortuous and has been worked out in only a very few cases so far. There are about 4000 inherited human genetic diseases. In only a few cases, such as muscular dystrophy and cystic fibrosis, has the gene and its protein been identified.

From the point of view of understanding development, mutations are a fundamental tool. For, because they cause alterations in development, they offer the possibility of identifying genes which play a normal role in development. Once a mutation that causes an abnormality is observed, it becomes possible to try and identify and isolate the gene. If the gene is isolated then it can be cloned—that is, large numbers of copies made—and then its nucleotide sequence determined. From the nucleotide sequence the amino acid sequence of the protein it codes for can be determined and this provides an excellent starting point for finding out the role of the protein in development. This approach has been very successful in the study of insect development (Chapter 7).

However, discovering the function of a gene is not at all easy. For example, there are mutations in mice which cause abnormalities in limb development, such as reduction in digit number, but there is no simple or direct way to isolate the genes which are responsible. It is rather like looking for a single straw in a haystack. However, in 1985, a fortuitous observation made in the Harvard laboratory of the distinguished geneticist, Phil Leder, seemed to offer a way of identifying one of the genes. Leder and his colleagues were working on tumour viruses that cause cancer, when they noticed that one of the mice infected with the virus had a limb deformity. They correctly assumed that the viral DNA had been inserted into the DNA of a normal gene involved in limb development and so had caused a mutation. But they now had a great advantage because they knew that their viral DNA was in the gene and so they could use this knowledge to fish out the gene that clearly plays some role in limb development. This they did, and they then determined the nucleotide sequence of the gene, and thus the protein it codes for. Now, in most cases when a new protein is found from a gene sequence, some idea of its function can be inferred from the similarity of its amino acid sequence, to those of other proteins whose function is known. But not in this case. It was like no other protein that anyone had described before, and so there was no clue to its function. Thus at present they have a gene that controls limb development, but they have little idea what its role is. Their approach now is to try and find out where and when the protein is made during limb development, in the hope that this will provide some insight as to how the gene works.

CONTROL OF GENE ACTIVITY

Control of protein synthesis is the central issue in cell differentiation and development. More accurately it is the control of the 'luxury' molecules—the molecules that make cells different from one another, rather than the control of the 'housekeeping' proteins—which is the key. If one thinks of the DNA codes for proteins as records in a juke-box, the problem is to understand why one disc, say the haemoglobin theme, is played only in red blood cells while another, the albumin theme, is played only in liver cells. What presses the right button in different cells?

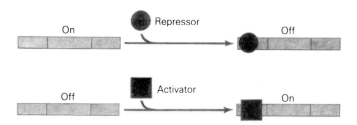

Control of protein synthesis can occur at several different points in the sequence of steps that leads from the DNA code to a fully formed protein. The first step, transcription, is considered by far the most important. Control of transcription is done in two main ways. Proteins can bind to sites on the DNA at the beginning of the gene—known as the promoter—and so initiate transcription; on the other hand, there are proteins that can block transcription by binding near the promoter and so prevent its transcription. In a sense a great deal of development is about transcription factors, for these factors regulate gene activity and so the state of the cell.

The molecules that control transcription are themselves proteins which are coded for by other genes. So, what determines whether these genes are transcribed? There is, in fact, a complex network of gene interactions, the product of one gene controlling the activity of others which involves an intimate relationship between nucleus and cytoplasm, because cytoplasmic signals play a crucial role in controlling transcription and thus determining whether genes are on or off.

The control of transcription—turning on and off of genes—and so controlling the synthesis of specific proteins, involves signal molecules that enter the cell nucleus from the cytoplasm. Consider the following experiment. The red blood cell of chickens (unlike our own red blood cell) has a nucleus. It is, however, almost as if it did not have one. The nucleus is quite inactive—no genes are on, and no messenger-RNA for new proteins is being made, and there is no protein synthesis. The human cancer cell could not be more different. Many genes are transcribed and many proteins synthesized. The question is what will happen to the red blood cell nucleus if it is placed in cancer cell cytoplasm? The particular cancer cell used in the experiment is the HeLa cell line which can be grown in large quantities in flasks. (Helen Lane, from whose tumour the cells were isolated many years ago, gives the cell line its falsely exotic-sounding name.)

The way the red blood cell nucleus is put into the HeLa cell cytoplasm is by fusing the two cells. As the red blood cell has little cytoplasm, the effect is that both nuclei become surrounded by HeLa cytoplasm. The HeLa cell nucleus continues to behave as before. But there is a dramatic change in the chick red blood cell nucleus. It gets larger, and dormant chick genes become reactivated within a few days. For example, new molecules characteristic of the chick appear on the cell surface. This implies that signals from HeLa cell cytoplasm have entered the chick nucleus and brought about this gene activation.

Another example illustrating the importance of cytoplasmic factors in continually controlling gene activity comes from fusing mature muscle cells with cells of a completely different type. Mature striated muscle cells make muscle-specific proteins and when such a cell is fused with a variety of other cell types which include liver and cartilage cells, then muscle-specific genes are activated in these cells leading to the production of muscle-specific proteins. Identification of the new muscle proteins was made possible by fusing human cells with mouse muscle cells; the human muscle proteins are similar but distinct from those made by mice.

There can be no doubt, from these experiments, of the importance of factors in the cytoplasm controlling gene action. But how far does it go? Is the DNA in the nucleus really that passive, behaving like a juke-box from which any cytoplasm will select the correct themes? Could it be that the cytoplasm completely controls gene action? A famous experiment that involves transplanting nuclei into the egg provides an answer, and more.

TRANSPLANTATION OF A NUCLEUS

It is possible to remove the nucleus from the egg of a toad. The toad egg is large and yolky and the nucleus floats in the cytoplasm near the top of the egg just beneath the surface. If the top of the egg is exposed to a sufficient dose of ultraviolet irradiation the nucleus is inactivated and is functionally useless while the cytoplasm is unaffected. The egg, effectively, has no nucleus. It is now possible to introduce another nucleus into this egg. The English embryologist, John Gurdon, in an extensive series of experiments performed during the 1960s, showed that if he transplanted the nucleus from one of the cells of the blastula of early toad embryo it could function as if it were the toad egg's nucleus. In a large number of cases the egg would develop quite normally into a swimming tadpole and then an adult toad. Impressive though this result is, the transplanted nucleus had come from an early embryonic stage. Could the same result be obtained by transplanting a nucleus from well-differentiated cells?

Gurdon transplanted a nucleus from the cells lining the gut of a swimming tadpole into the 'enucleated' egg. He observed normal development of the egg even though the number of times this occurred was much reduced. He even transplanted nuclei from adult skin cells in culture and again found that normal development could take place.

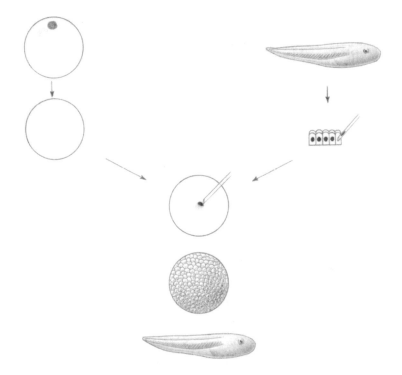

This shows conclusively that the egg cytoplasm could completely alter the gene activity in the transplanted nucleus, because the genes active in gut and skin cell nuclei are quite different from the genes active in early development.

Since the nucleus of a cell in the gut could give rise to a normal toad it is clear that no genetic information had been lost during the development of the gut, and the same holds for the skin nucleus. The important conclusion is that all the cells in the body contain the same genetic information; what makes the cells different is how that genetic information is used. Since the nucleus of each cell contains the same genetic information, it is the reciprocal communication between nucleus and cytoplasm during development that determines which proteins are made.

There is a further implication of the experiment. Nuclear transplantation provides a way of creating clones of identical toads. Since the nuclei from one animal contain identical genetic information, all the animals that develop from grafting nuclei taken from one toad into enucleated eggs will be identical. Gurdon has, on occasion, made such a clone of 10 toads. The members of this family are genetically identical and this was confirmed by making skin grafts between them and finding that they were not rejected.

One should not, however, think that the nucleus of any cell will support development if transplanted into the egg. The nucleus must be taken from cells that are multiplying. Nuclei taken from brain cells, which never divide, will not support development. There is also another limitaton which is seldom mentioned: the toads that develop from nuclei taken from mature cells, like gut epithelium or skin, are infertile. The reason is not clear, but must be the result of some subtle change in the genetic information in the nucleus.

One of the fear-filled fantasies that many people have about modern biology—genetic engineering—is the possibility of cloning identical people. This idea comes from false reports that cloning of a human has already been done. Some millionaire or other, it is sometimes claimed, has had himself cloned in some secret clinic in Switzerland (or South America) by taking the nuclei from some of his own cells. This is simply nonsense. No mammal has been, or at present could be cloned in the way described for toads. Quite the contrary, all the evidence from experiments with mice, which have been intensively studied, shows that it is not even possible for the nucleus of a very early embryonic mouse cell to substitute for the egg nucleus. By the four- to eight-cell stage of development the DNA in the nucleus has already undergone some apparently irreversible changes. This does not necessarily mean that the genetic information has been lost, but only that there may have been some chemical modification which makes crucial genes inaccessible. But it does mean that, at present, cloning of mammals is not possible by nuclear transplantation.

While the nuclear transplantation and other experiments make it clear that all nuclei contain the same genetic information, there are exceptions. In certain worms, for example, there is loss of DNA from some cells, only the future germ cells retaining the full complement of DNA. Also, some cells, like the mammalian red blood cell, have no DNA at all, the nucleus being expelled during differentiation.

GENE SWITCHES

The switching on and off of genes in different cells is fundamental to the programme of development. It underlies and initiates the changes seen in development. It is also among our deeply held convictions that gene switching is the basis of both cell memory and change. Liver cells remain like their parents because the same genes are switched on. Cells in the leg are different from cells in the arm because the pattern of activity of the genes is different. Of course, switching on and off of genes does not in itself result in developmental change; switching genes on and off acts by changing which proteins are made and this in turn alters cell behaviour by pathways that, to repeat, may be quite tortuous.

Gene action is controlled at the level of transcription, and whether a gene is on or off is whether or not it is transcribed. Control of transcription is largely related to the state of the promoter at the beginning of the gene sequence. That the control of gene activity that characterizes a particular cell type is due to specific factors in the cell acting on the promoter, comes from experiments in which the promoter elements are switched around. Genetic engineering makes it possible to join the promoter of one gene to that of another. When this DNA construct is incorporated into the DNA of the developing embryo the protein will be made in those cells where the promoter is activated, even if the protein is quite inappropriate to the cell.

Growth hormone is a protein made in the pituitary gland at the base of the brain. It is essential for normal growth and is secreted by the pituitary and carried in the blood to all parts of the body (Chapter 10). The gene for growth hormone has been isolated and the controlling promoter region identified. Genetic engineering makes it possible to cut and splice DNA. So, it was possible to replace the promoter for the growth hormone gene with the promoter for a protein which binds metal ions in the blood: this promoter is activated when traces of metal are present. The new construct—the growth hormone gene joined to a promoter that is turned on by traces of metal—was injected with a fine pipette into the nucleus of a fertilized mouse egg. Many copies were

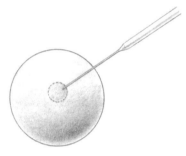

injected and some of them inserted, almost randomly, into the DNA of the nucleus of a fertilized mouse egg. The egg was then allowed to develop and all the cells of the embryo now contained multiple copies of the growth hormone gene and its new promoter. Shortly after birth, the mouse was fed water containing traces of metal, so activating the injected construct; the result was a mouse almost twice the size of a normal mouse, and was affectionately labelled 'Supermouse'. The growth hormone gene had been turned on in all the cells throughout the body and the increased production of the growth hormone had resulted in increased growth.

Supermouse is known as a transgenic mouse. It has been given new genetic information by injecting DNA into the nucleus and this will be inherited by all the cells in the body and passed on to future generations via the germ cells. Transgenic mice provide a particularly powerful way of studying how genes affect development and how they are turned on and off.

There is another experiment which demonstrates the presence of factors in cells that turn on specific genes. It makes use, again, of the growth hormone gene, and another gene which codes for an enzyme elastase, which is made in the pancreas and enters the stomach where it breaks down elastic tissue. The promoter of the elastase gene was joined to the growth hormone gene and this new construct injected into the nucleus of the fertilized mouse egg. It was now found that growth hormone was made in the pancreas. Specific promoters are activated in specific cells. It is as if the promoter regions for luxury proteins contain a list of 'addresses' at which the gene should be active.

The state of a cell can, in principle, be described by its genetic activity—which genes are switched on and off. Development can in this sense be viewed in terms of the changing networks of gene activity in dfferent cells. Activation of one gene can lead to the synthesis of a protein which activates some genes and inhibits others, which in turn may control yet other genes. It is not yet clear just how complex these networks of gene activity are. One possibility is that there are master genes whose products control the activity of many others. There is, perhaps, a feeling that there are so many factors, so many interactions, that the whole system is going around in a circle, almost out of control. This is not the case: rather, this feeling reflects much of our ignorance as to what, in detail, is going on inside the cell.

Ultimately, we want to link gene action with pattern and form. Knowing how genes are turned on and off is an important step towards understanding the internal programme of the cell and how this is influenced by signals from other cells. During development a number of different cell types are generated, each with their characteristic 'luxury' proteins. How is this diversity achieved?

CELL DIVERSITY AND DIFFERENTIATION

THE SINGLE cell, the human egg, gives rise to about 350 different cell types, blood, muscle, skin, and so on. This diversification requires that the genes coding for the 'luxury' proteins in these cell types are switched on in the appropriate cell. A useful image of cell diversification is of an undulating landscape in which a ball rolls down pathways that branch, an image that the British embryologist Conrad Waddington, writing in 1940, called the epigenetic landscape. At many branch points there may be just two new tracks, while at others there may be more. The tracks can be thought of as patterns of gene activity and the ball as a developing cell. Which particular pathway a cell follows is usually controlled by extracellular signals acting at the branch points. Cells can rarely reverse tracks, but some treatments can push the cell from one track to an adjacent one. Overall, the image gives a feel of the nature of diversification during development, since as a cell proceeds down a pathway and takes one of the branches, other pathways cease to be open to it. With time, the cells proceed further and further down the landscape and become more and more distant from each other, reflecting different patterns of gene activity.

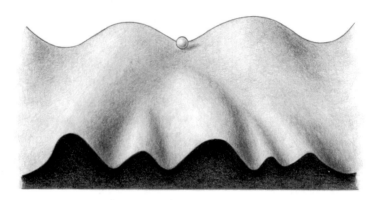

The epigenetic landscape is a representation of changing gene activity. Every cell has the same genetic constitution and some genes are switched on and off as the cell develops and rolls down the landscape. As the cell proceeds down a pathway genes are switched on and off reflecting an internal programme, for external signals only act at the branch points, directing the cell along one or other of the paths open to it. Knowledge of the pattern of gene activity is still very limited. In general, the picture is still painted with a very broad brush and later stages are better characterized than early ones.

The final pathway has no branch points and is essentially maturation. For example, the maturation of the red blood cell involves major structural changes and the loss of all its genetic information since the nucleus is expelled. Before this stage a large amount of messenger-RNA for the protein haemoglobin is made and stored in the cytoplasm, so that when the nucleus is lost synthesis of haemoglobin continues. But even that stops and there is no protein synthesis at all. The red blood cell is finally a small bag containing haemoglobin molecules for transporting oxygen. Muscle cells, too, go through a series of characteristic changes as they mature. They take on an elongated shape and fuse with other muscle cells to give a multinucleate muscle fibre. New proteins are now synthesized which provide the muscle's contractile machinery and it becomes organized in the cell into a highly ordered array of filaments which gives muscle its striated appearance. But what commits the muscle to this final pathway and what ensures that all the necessary genes are turned on at the same time?

In the case of muscle it seems that there is a master gene controlling the expression of all the main genes, known as myogenin and it is always switched on in mature muscle cells. Its special property is that it can activate most of the genes that are required in the mature muscle. The myogenin gene has been isolated and can be introduced into non-muscle cells where it becomes integrated into their DNA. The effect of introducing just this one gene is to switch off some genes and to set in train a sequence of gene activation so that the cell develops into a muscle cell with all the appropriate proteins—master gene is an appropriate description. But whether this is peculiar to muscle, or whether other master genes will be found, remains to be discovered.

One can imagine how an external signal could activate a gene like myogenin and so guide the cell down the muscle pathway. Such branch points are crucial in the development of diversity and are particularly clear in the development of blood cells, of which the red blood cell is only one representative.

THE BLOOD LINEAGE

Red blood cells in adult mammals only have a limited life span—about 120 days in humans—and are continually being produced. There are, in addition to the red blood cells, which are the major population, seven other kinds of blood cells: they thus provide a vital and valuable system for studying cell diversification. Among the white cells are scavenger cells collectively known as granulocytes; lymphocytes, which are responsible for the immune response; macrophages, which are also scavengers; and a type of cell which makes the platelets involved in the clotting of blood when an injury to a vessel occurs. The

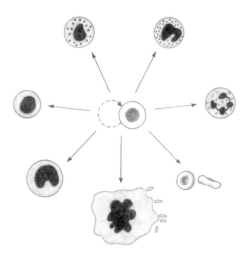

generation of blood cells takes place primarily in the bone marrow and spleen.

All the cells in the blood come, remarkably, from just one special progenitor cell—the multipotential stem cell. The essential nature of a stem cell is that it is self-renewing and, as its name implies, the source of other cells. When the stem cell divides one of the two daughter cells may go on to give rise to other types of cell, whereas the other

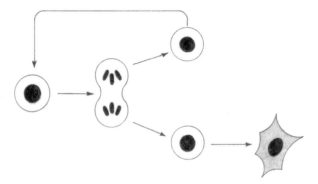

daughter cell remains a stem cell, capable of dividing again and always giving one daughter to diversification. Thus a characteristic feature of stem cells is this asymmetry; one daughter keeping the stem cell character, the other proceeding along a pathway of diversification. In principle, because stem cells are self-renewing, they are, unlike the cells they generate, immortal.

The evidence that there is just one kind of cell which generates all the blood cell types—a multipotential stem cell—comes from destroying the stem cells of mice and then replacing them with just stem cells. X-irradiation destroys the stem cells and the resulting depletion of blood cells—no more are made—is the basis of radiation sickness, and leads to death. Rescue of the animal is possible if the stem cells are replaced. Injecting as few as 20 multipotential stem cells into a lethally irradiated mouse will suffice. After many years of patient work, it has been possible not only to isolate the multipotential stem cells but provide conditions such that they will differentiate into the different kinds of blood cell in culture.

The differentiation of the different kinds of blood cells can be thought of in terms of the branching pathway model. Each branch point can be thought of as occurring at a cell division and so the branching pattern is also a cell lineage which starts with the multipotential stem cell. The key feature in the programme of differentiation of the stem cell is one of asymmetry. When the stem cell divides the two daughter cells behave differently. Is this asymmetric behaviour an intrinsic property of the stem cell, or are there environmental signals which control the different pathways of the daughter cells?

Since the various types of blood cell will develop in culture from the multipotential stem cell, it suggests that the programme of diversification may be intrinsic to the cells and not due to external factors. In culture, the cells are not exposed to the factors that cells in the spleen or marrow might secrete or other factors circulating in the blood. On the other hand, there is extensive evidence for specific factors which can control the pathways of differentiation. At one stage in the lineage, for example, there is a cell which can produce either granulocytes or macrophages, and the decision to follow a macrophage or granulocyte pathway can be controlled by the concentration of a protein factor which, if the concentration is high enough, will direct differentiation exclusively along the macrophage pathway.

The detailed lineage of blood cells is not fully worked out but it is quite clear that there is, with successive cell divisions, an irreversible commitment to one or other general pathway. With divisions the multipotentiality of the ancestral stem cell becomes progressively reduced: the choices open to the cells become fewer and fewer. As the cells proceed from the stem cell to the various mature blood cell types they divide many times. While some of these divisions represent a branch point, many of the divisions are only proliferative, and serve to amplify the number of cells at that particular stage in the lineage. Very large numbers of blood cells are required—human red blood cells only last a few months—and there are few stem cells, so an enormous amplification is required. Factors such as the protein erythropoietin can stimulate red blood cell amplification and there is an increased production of erythropoietin when there is a shortage of oxygen, caused, for example, by going to a higher altitude.

DYNAMIC LININGS

Immortal stem cells and cell diversification are also fundamental to both the internal and external linings of our body. Both our skin and the lining of our gut are always being replaced. Dead skin cells are continuously falling off and contribute to the dust in our houses. The lining of our gut also sloughs off continually. These cells must be

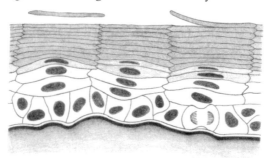

replaced and involve a process similar to that of blood cells. Beneath the surface layer of the skin are several further layers of cells. At the base of the skin there is a layer of dividing cells which include stem cells, and these provide the cells which move upwards to the surface. When cells leave the basal layer they cease to divide and begin to mature into the tough dead cells that cover the surface. As they move upwards they lose their nuclei and synthesize the special proteins like keratin that give the skin its protective toughness. The overall thickness of the skin is controlled by the rate of division of the basal cells and removal of cells from the outer layer somehow increases the rate of cell multiplication to restore the thickness; perhaps the dying outer layers produce an inhibitor of cell proliferation so that if the number of surface cells is reduced inhibitor concentration decreases and more cells are produced.

The life span of a human skin cell from the time it is born until the time it is shed is just a few weeks. Replacement is dependent on the continued multiplication of immortal stem cells. It is likely that the stem cells maintain their characteristic state by virtue of their position in the basal layer where they may be under the influence of a deeper layer of cells. As with blood, X-irradiation kills off the stem cells, so that when cells are lost from the surface there are no new cells to replace them. The underlying tissues thus become exposed giving the skin the appearance of having the upper layers destroyed as in a burn.

Surface cells are also continually lost from the gut. This lining is highly folded to increase the area for the absorption of nutrients. The cells are lost from the tips of the folds and the stem cells are located at the very base of the folds. As cells move upwards from the base, they continue to divide, but, unlike the stem cells, they have lost their immortality and they stop dividing when they approach the surface where they are sloughed off. Again, just as it does in the skin, X-irradiation kills off the stem cells and this soon has dire effects on the surface of the gut, giving rise to one of the major causes of radiation sickness.

In the gut, as in the skin, the properties of the stem cells seem to be maintained by their basal position. Once the cells leave this position they cease to be stem cells and are committed to a pathway leading to maturation and death.

THE DIVERSE NEURAL CREST

One group of cells shows a particular diversity in its developmental pathway. When the neural tube closes (Chapter 2), a group of cells detaches itself at the site of fusion and migrates to many parts of the body and differentiates into a wide variety of cell types. This has already been mentioned in the discussion of cell migration (Chapter 2).

It has been possible to map in detail the migration and fate of the neural crest cells in the chick because of a fortuitous discovery made in 1968 by the French embryologist Nicole Le Douarin. She noticed that the nuclei of the cells of the quail embryo looked slightly different from those of the chick embryo. Because transplanted quail cells will behave normally in chick embryos, she realized she had an invaluable natural marker. By grafting an early neural tube from the quail embryo into a similar position in the chick embryo—the corresponding portion of the chick embryo having been removed—Le Douarin could follow the fate of the quail neural crest. Her work showed that the cells migrate to many different sites in the embryo developing into the skeletal elements of the head, all the pigment cells in the body, most of the nerves of the involuntary nervous systems, sensory nerves, and a variety of glands.

Neural crest diversification presents a special problem. Not only must the cells differentiate into a wide variety of cell types such as cartilage and neurons, but they have to migrate long distances to very special sites in the embryo. There are several ways this could be achieved. The cells could be multipotential, migrate to all the sites and only once they had arrived would they, due to local signals, be directed along the correct developmental pathway. Or, they could have their developmental pathway specified before they begin migrating and then would migrate to the correct sites. Cells programmed to develop as cartilage would, for example, migrate only to the head region.

There is yet a third possibility. The migrating cells may be made up of a mixture of all the different cell types in immature form, that go to all the sites and a particular type survives only if it arrives at an appropriate site—a sort of cell selection. Embryos have their own logic and all too often eschew tidiness; there is an element of all three mechanisms involved.

The experimental approach is to use quail neural crest, and instead of grafting the quail cells to the corresponding position in the chick, they are grafted to an inappropriate site. For example, head neural crest may be grafted to a more posterior position along the embryo and posterior neural crest grafted to the head region. The result shows that head neural crest is from the very beginning different from neural crest in other regions, and any other neural crest placed in the head leads to abnormal head development. Even though head neural crest is essential for the head it can satisfactorily replace trunk crest, suggesting its cells are multipotential and that their differentiation is determined by where they end up. Evidence for the third possibility, cell selection, comes from regrafting cells that have already migrated and begun to differentiate, to the site of neural crest formation in a younger embryo. These older cells undergo a second migration and give rise to a variety of cell types quite alien to the site at which they had arrived in their first migration, suggesting that there is a mixed population of cells at each site at the end of migration and the conditions at each site favour the growth and differentiation of specific members of the mixed population; the others fail to flourish and presumably die.

While all the neural crest cells are not completely multipotent their fate is far from fixed as they begin to migrate. During this migration they receive signals from the surrounding tissues which directs them along the appropriate developmental pathway. In this respect they are similar to the developing blood cells whose diversity is also controlled by a sequence of signals from the surrounding tissues.

THE GLIAL PROGRAMME

Most mature cells have characters, such as a distinctive shape, which make them easy to identify. There is no difficulty in recognizing a red blood cell, a muscle cell, or a nerve cell. But in following embryonic cell lineages of blood or neural crest the early cells are bewilderingly similar; there are no obvious features which can be used to distinguish, say, early red blood cells from white cells. Since the commitment to a particular pathway occurs long before clear changes in shape emerge this can be a major obstacle to studying differentiation. Major

advances in overcoming this difficulty have been made by using antibodies to distinguish between cells that seem very similar. Antibodies
are proteins that are specific in binding to different molecules; thus
antibodies can be made which recognize only one kind of molecule on
the cell surface. If that surface molecule is unique to one kind of cell,
then by using a fluorescent antibody, cells can be recognized under the
microscope by the presence of fluorescence on the cell. Antibodies
provide highly discriminating probes for distinguishing cells which
contain any of the many different molecules against which the antibodies can now be made.

The antibody technique was essential for teasing out the roles
played by external signals and internal factors in the differentiation of
glia—the supporting cells of the nervous system—so that is now one
of the systems we understand best. In the optic nerve of the rat—the
nerve that joins the eye to the brain—there are only three kinds of glial
cells. Two of these are astrocytes, which are supporting cells occupy-
ing the spaces around the nerve cells; the third kind are oligodendrocytes which wrap closely around the nerve, providing an insulating

Astrocyte

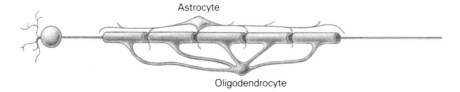

Oligodendrocyte

layer which speeds conduction of the nerve impulse. One kind of
astrocyte and the oligodendrocytes are derived from a common dividing progenitor cell which can differentiate into either cell type. Oligodendrocytes appear at about the time of birth and the astrocyte a week
later. How does the progenitor cell 'know' when to make oligodendrocytes and when to make astrocytes?

The progenitor cells respond to factors in their environment. In the
presence of a growth factor the cells produced by the progenitor cell
go through about eight cell divisions before differentiating into
oligodendrocytes. The switch to making astrocytes only occurs when
another factor, which is made by the optic nerve, is present. An

impressive demonstration of the role of this factor is its effect on a single progenitor cell which can be followed under a microscope. In the absence of the factor the cell becomes an oligodendrocyte; in the presence of the factor, an astrocyte.

This system shows the role of external and internal factors in controlling the pathways of differentiation of cells. The factor made by the other astrocyte cells is necessary for the progenitor cell to enter a programme of eight cell divisions which has an almost dance-like character. Only when that programme of cell divisions has been carried out do oligodendrocytes differentiate. The progenitor cell now receives a signal from the adjacent optic nerve which switches its pathway of differentiation from an oligodendrocyte to that of an astrocyte.

CHIMAERAS

Chimaeras have always been a source of some horror. Mythology is full of monsters, half one beast, half another; the head like a bull, the tail like a snake. Yet all female mammals are chimaeras; half the cells have one X chromosome active, and in the other half the other X chromosome is active. Females have two X chromosomes and males just one (Chapter 9) and this imbalance in chromosome number is developmentally unacceptable. Embryonic development is very sensitive to having just the right number of genes and so in females one of the X chromosomes is inactivated early in development, leaving just one functional X chromosome in each cell. Inactivation of the X chromosome is random so that the early embryo is made up of a mixture of cells in which one or other of the X chromosomes is inactivated. The embryo is thus a chimaera, a mosaic of two different kinds of cells. A visible example of X chromosome inactivation is the tortoiseshell cat. Genes controlling coat pattern are carried on the X chromosome and if one X carries a gene that gives colour and the other X carries an inactive gene, the different patches of tissue containing the different inactivated X chromosomes show up as patches of different colours.

The inactivation of the X chromosome is a good model for cell 'memory' since it remains inactivated through cell growth and cell division. It is thought that the mechanism whereby all the genes in the chromosome are inactivated is a chemical modification of the DNA.

A quite different way of creating a chimaera is to fuse two early mouse embryos. This can be done just by pushing them together. Their cells then intermingle and a normal mouse can develop.

However, not quite normal, since the mouse will have four parents if the original two embryos came from quite different mothers and fathers. If, in addition, the one embryo carried a gene for giving the mouse a black coat, and the other lacked such a gene, thus having a white coat, the result is a striped mouse.

REGULATORY GENES

Having drawn a distinction between 'luxury' proteins and 'house-keeping' ones (Chapter 5), and, by implication, the genes that code for them, a further distinction is helpful: the distinction between regulatory and structural genes.

Structural genes code for luxury proteins that play an important role in the life of the cells—haemoglobin in red blood cells, contractile proteins in muscle. The products of regulatory genes are only involved in regulating other genes and their products do not serve any other function. One such gene is myogenin which activates a set of genes in muscle development. Regulatory genes are very important in pattern formation. As pointed out in Chapter 3 the difference between ourselves and chimpanzees does not lie within the cell types but in their spatial organization. This view has a strong support from molecular studies which show that the proteins from chimpanzees and man that characterize cell types are very similar. The differences between chimpanzees and us cannot be accounted for by differences in these proteins. Rather, the differences must be sought in the regulatory genes and proteins that control spatial organization. The best chance of our understanding this link comes from studies of the fruit-fly.

GENES AND FLIES

T HOMAS HUNT MORGAN would have been particularly pleased with the impact that fly genetics has had on understanding development. Morgan, an American, initially worked on development and regeneration at the beginning of the century and was the first to clearly put forward ideas on how gradients could control patterning. The story is that he decided that the problem of embryonic development was just too difficult so he turned to genetics and used the fruit-fly as his model. His studies revolutionized genetics. He could not have foreseen how studies of the fly have illuminated how genes control early patterning in the embryo so that a link between gradients and genes would be established.

The combination of genetic and embryological studies have identified the regulatory genes that control patterning in the early embryo of the fruit-fly *Drosophila*. These studies have also led to the discovery of a portion of a gene, known as the homeobox, which has enabled regulatory genes controlling pattern to be identified in many other animals including humans.

Unlike the eggs of many other animals, the egg of the fruit-fly is not round but an elongated cylinder. The sperm enters from one end and its nucleus fuses with that of the egg, to give a single nucleus. Again, most unusually, the egg does not undergo proper cleavage but, instead, the nucleus alone undergoes a series of divisions, about one every 8 minutes. This results, within two to three hours, in some 5000 nuclei floating in a common cytoplasm. Only then does this single-cell embryo, with many nuclei, begin to develop walls between the nuclei, and the system becomes a more respectable multicellular embryo.

Gastrulation takes place, and after 24 hours the embryo has developed into a segmented feeding larva. The larva gives only a hint of the fly which will eventually emerge, the similarity being that both larva and fly are segmented. There is a head at one end of the larva, or, essentially, a structure for taking in food. Then come three thoracic segments, followed by eight abdominal segments, ending in a tail-like structure. The difference between the thoracic and abdominal structures is subtle and can only be recognized by the arrangement of bristles. Drosophilists have great skill in recognizing and interpreting these patterns in abnormal animals. The larva feeds and then forms a pupa, undergoes metamorphosis, and out pops a fly.

The adult fly can also be thought of as being made up of a number of repeated structures, the segments. Behind the head are three thoracic segments followed by six abdominal segments. The thoracic segments carry the wings and legs. Segments are fundamental to the fly's structure and even the head is really made up of segments that have fused. The segments are both very similar and very different and understanding this apparent paradox lies at the heart of fly development.

SEGMENTING AND PATTERNING THE EMBRYO

Segments are fundamental to the fruit-fly's development. A recent discovery of major importance has shown that the boundaries of the segments are already present in the embryo before cell walls have surrounded the nuclei. This was quite unexpected as no one had anticipated that the basic segment pattern would be laid down in what is essentially a single cell. Certain mutations affected the pattern of segmentation in the larva; for example, mutations in the so-called pair-rule genes *hairy* and *fushi terazu* resulted in every second segment being missing. (The naming of genes is, to put it mildly, often idiosyncratic.) It was thus natural to try and find out when and where these genes were expressed in the early embryo. The surprise was that the genes became active in a stripe-like pattern before cell walls form and while the embryo is still a single cell. The genes become active in lines of nuclei at right angles to the long axis of the egg giving a beautiful stripe-like pattern which defines segment boundaries. Initially, there are seven stripes about four cells wide which then become subdivided giving the 14 stripes of the embryo which will become the definitive segments.

The striped pattern has, quite reasonably, led quite a few workers to speculate that the stripes are generated by a mechanism that generates a repeated pattern, like the peaks in a wave in a reaction-diffusion system (Chapter 3). But the fly is full of surprises. For all the evidence points to each stripe being specified separately and independently of the others by the activity of genes that act even earlier than the pair-rule genes. These are involved in setting up the initial pattern along the main axis of the embryo. One of the most important is the gene *bicoid* which is involved in patterning the anterior end of the embryo.

The polarity of the embryo, that is which end will become the head end and which the tail, is built into the egg when it is developing in the mother fly. Special cytoplasm is located at the future anterior end. In this special cytoplasm is the message for synthesizing the protein coded for by the *bicoid* gene. When the egg is laid the *bicoid* protein begins to be synthesized at the anterior end and diffuses along the egg setting up a beautiful concentration gradient whose high point is at the front end. Beautiful because it is the first unequivocal demonstration in any developmental system of a gradient that controls patterning. After years of inferring the existence of such gradients here, at last, is

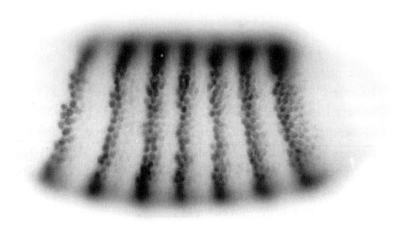

Activity of a pair-rule gene in the early Drosophila *embryo*

the real thing! This gradient controls the position of the boundary between the head and thorax and also activates another gene at a specific position along the embryo.

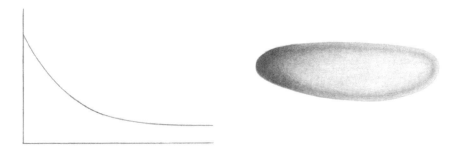

The role of the *bicoid* gene protein in normal development is demonstrated by the effect of mutations in the *bicoid* gene. In the eggs of flies that lack a normal *bicoid* gene there will be no *bicoid* proteins at the anterior end and the embryo will develop into a larva lacking both head and thorax. The abnormal eggs can be rescued by injecting anterior cytoplasm into the anterior region of the egg: by restoring the gradient normal development will now take place. If, however, the anterior cytoplasm is injected into the middle of the egg the head will now develop in the middle of the embryo, which is consistent with the special cytoplasm defining the anterior end.

The gene, *bicoid*, is one of the first to be activated during patterning of the egg. The *bicoid* protein activates other genes and there is a further set of genes which is involved in patterning the posterior end of the embryo. As a result of such gene actions there develops, in the embryo, a pattern of proteins of varying concentrations which causes the activation of the pair-rule genes at specific locations and so leads to the stripe pattern and then to defining the segment boundaries. The promoter region of the pair-rule genes can recognize and respond to particular patterns of protein concentration, thus enabling the stripes to be specified individually. Each stripe has its own 'address'. While the details are quite well understood, it would be misleading to suggest that this sequence has an elegant simplicity.

Embryos are not one-dimensional and at right angles to the main body axis is the dorso-ventral axis whose pattern is controlled by a set of genes quite different from the antero-posterior axis. Again, a gradient in a protein is set up, but this time a rather different mechanism is used to establish it. Rather than special cytoplasm being laid down on the ventral side, there is a change in the cell membrane on the ventral surface, which sets in train a series of reactions leading to a gradient in the protein of the *dorsal* gene. This gradient in the *dorsal* protein, with its high point ventrally, is confined to the cell nuclei, and directly controls the genes causing patterning along the axis. The

gradient in the product of the *dorsal* gene then directly controls what the cells will do. At a high concentration the cells make muscle, at an intermediate concentration nerve cells, and at the lower concentration surface cells. It is probably the clearest demonstration of a gradient in positional information controlling development.

Our current understanding of the early development of *Drosophila* in terms of gene activities is a brilliant success story. It provides the best example of a developmental programme. Much of the credit must go to Christiane Nüsslein-Volhard who works in Tübingen, not far from Freiburg where Spemann discovered the organizer (Chapter 3). Her success comes in part from her decision to invest a great deal of effort in identifying the genes that control early development. Most of these genes, which number about 50, were identified and this opened up the possibility of finding out how they interacted and controlled early development. It was a brave undertaking and even she could not have predicted how successful the outcome would be.

The action and interaction of the early pattern genes result in the early embryo being divided into segments. They also provide a basis for making the segments different from each other, so each has a unique identity. Segment identity is finally specified by another set of regulatory genes, the so-called homeotic genes. The term 'homeosis' was coined by the English geneticist, William Bateson, at the end of the last century to refer to the process whereby, on occasion, one structure is replaced by a different one.

HOMEOTIC GENES

There is a mutant gene *aristapaedia* which results in the fruit-fly having a leg growing out from the head in place of the antenna. This change in structure is due to a defect in just one single gene. There are

a number of mutations in genes which cause the replacement of one fly structure by another—these are the homeotic mutants. For example, other homeotic mutations can cause a leg to be replaced by an antenna, or an additional pair of wings to form, and yet others transform the eye into a wing. The normal function of these homeotic genes is to specify that structures develop in the right place; the mutations result in the structures developing in the wrong place.

A way of thinking about the normal function of homeotic genes is to regard them as providing each segment with a unique identity, not unlike a positional value. Different homeotic genes will be turned on in different segments and so can control both the nature of the segment and what structures, such as wings or legs, develop there. Justification for such a model is provided by completely deleting all those genes which are responsible for segment identity; the larva that develops from such embryos has all its segments the same. This also shows that the process of making segments and giving them a separate identity are different. However, while they can be independent they are also linked, for the set of gene actions during early development that lead to segment formation must also activate the homeotic genes in the correct segments. So homeotic gene expression is near the end of quite a long and complex chain of interactions in early development.

There is one further striking feature of those homeotic genes that specify the identity of the segments in the thorax and the abdomen: the order of the genes on the chromosome corresponds to the order in which they are activated along the body axis. The genes are together in a cluster and it is as if they are sequentially activated along the thorax and abdomen. It is an elegant surprise to find a correspondence between the spatial pattern of genes on a chromosome and the pattern of structures they specify.

One is always looking for a general principle to try and understand gene action and cell patterning, and gradients may provide one such unifying mechanism. But in many cases the role of gene action may be unexpected and apparently *ad hoc*. A striking case is that of the *nanos* gene whose protein product is present at the posterior end of the embryo. The gene is necessary for normal development of the abdominal segments but its role is very indirect. It turns out that *nanos* does not provide positional information, nor does it specify local cells, but its sole function is simply to turn off the activity of another gene in the abdominal region. There is no way such a function could have been predicted. It could not be deduced from general principles. Presumably its function is determined by the fly's evolutionary history in a way that, as yet, is not understood.

Since early *Drosophila* development is our best model for how genes control development it is worthwhile to recap some of the essential features of the system. Polarity must be established early on so as to set up the main axes and boundaries of the embryo. In the fly this is laid down in the mother when the egg is being made. Cytoplasm with special properties is located at the ends of the egg and, at the anterior end, a gradient providing positional information is set up. A separate gradient is set up along the dorso-ventral axis. The gradients along the body axes activate specific genes at specific concentrations and these genes in turn produce proteins which activate yet further genes. There is thus both a spatial and temporal hierarchy of gene activity; the early genes having their activity turned on in quite wide domains and controlling the activity of later genes, which have their activity restricted to more and more limited regions, leading to segment formation. Towards the bottom of the hierarchy the homeotic genes are turned on in specific segments and they can be considered to be at the top of a new hierarchy controlling activities within their segment.

ANTENNA AND LEG

The action of mutations in homeotic genes in changing, for example, the development of the antenna into that of a leg, is not understood. However, some of the principles involved at least define the problem in a rather novel way. It turns out that the leg and antenna are similar to the wing and leg in chick development (Chapter 4) in that each has the same positional information. The difference between antenna and leg is how the positional information is interpreted.

The experimental basis for this conclusion can be illustrated by using the analogy of flag development, but this time, using two flags, the French flag and the Stars and Stripes. Both only need red, white, and blue cells. If the positional values to make the two different flags are the same, and the differences lie in how these are interpreted, then there is a clear prediction. When, for example, a small piece of a developing Stars and Stripes is grafted into a developing French flag it should acquire new positional values and interpret them as if it were still in the Stars and Stripes. Thus, a piece of the Stars and Stripes placed in the top left-hand corner of the French flag will form a patch of stars; in the bottom right-hand corner a patch of stripes. The cells should develop according to their position and genetic constitution.

Exactly this type of graft was done in the antenna of a normal fruit-fly by 'grafting' in *aristapaedia* tissue. 'Grafting' in this case made use of a genetic technique but the effect is, in principle, identical. The striking result of this experiment is that the mutant cells develop according to their position along the antenna—just as in the flag model. If the cells are at the tip of the antenna they will develop into the claw of the leg; if they are at the other end they will form part of the femur. They develop according to their position and genetic constitution. The clear conclusion is that the positional information in the antenna and leg is the same and what makes an antenna or a leg is how this positional information is interpreted. This interpretation, in turn, is dependent on which homeotic genes have been turned on in particular segments.

THE EYE AND 'BRIDE' OF *SEVENLESS*

It would be misleading to think that all interactions in the development of the fly are similar to those in specifying the main axes. Many interactions seem very local and have no relationship to gradients. Rather, the problem is like specifying the position of each player in a rugby scrum. The development of the basic unit of the eye presents just such a problem.

Hundreds of repeats of a basic unit, which is relatively simple, make up the eye of the fly. This basic unit itself, the ommatidium, is made up of eight cells which respond to light and 12 supporting cells. The eight cells which respond to light—the photoreceptors—have a well-defined pattern in every ommatidium and develop in a particular order. Each photoreceptor occupies a different position and is named R1, R2, . . . R8. The eight cells are not related in any way by descent and the problem is what sort of interactions occur between the cells to generate this highly ordered rearrangement.

One class of model suggests that the pattern is generated by a very specific set of signals between cells. What would be required is that highly local and specific signals from early developing cells should tell later developing cells what to do. The signals would be expressed at very short range, perhaps only being transmitted between cells where their membranes are in contact.

Mutants have begun to provide clues to the interactions involved. One of the most interesting is the mutant gene *sevenless* which results in the failure of just one of the eight cells, R7, to develop. The failure of R7 to develop turns out to be due, not to the absence of a signal, but to the absence of something in the cell which can respond to the signal. It seems very likely that the signal necessary for R7 to develop is made by one of the other cells R8. If the *sevenless* gene codes for a protein that receives and responds to the signal, it is not unreasonable that the gene that codes for the signal, should be referred to as the 'bride' of *sevenless*; for their conjunction, for R7 at least, is a happy event. Such a gene has indeed been identified.

The development of the ommatidium is an excellent example of cell interactions and cell responses resulting in a small and highly ordered pattern. There are high hopes that quite soon all the signal and response elements will be identified. When they are it will provide a clear idea of just how many signals are involved in specifying a small but complex pattern, and whether there are any rules. Already there are indications of similarity in the signal molecules to those involved in quite different systems, such as factors which promote growth of cells in vertebrates.

THE HOMEOBOX

The success in understanding early fruit-fly development relies very heavily on the techniques of molecular biology which enable the genes to be isolated and studied in detail. Having isolated the gene, its DNA can be sequenced, that is, the four-letter nucleotide sequence determined, and so the protein that it codes for, worked out. A totally unexpected feature emerged when the sequences of some of the genes controlling early fly development were compared. A number of the genes contained a similar short sequence known as the homeobox which is about 180 nucleotides long, while the rest of the genes' sequence was quite different. The presence of a sequence of nucleotides which is very similar raises the question as to why this sequence should, as it were, have been highly conserved during evolution? It strongly suggests that it plays a similar and important role in all the genes in which it is present. All the evidence now suggests that the homeobox codes for that part of the protein which binds to DNA and is thus involved in the control of the activity of other genes. Many developmental biologists still cling to the idea that there are general principles. Because the homeobox was so highly conserved in genes that controlled patterning in insect development, the hope was that the homeobox would be found in the genes of other organisms.

That optimism was justified, for homeobox-containing genes have been found in worms, sea-urchins, chickens, mice, and humans. Moreover, these homeobox-containing genes seem to play a fundamental role in the development of these organisms. There is the exciting possibility that they are the molecular markers of position. For example, the pattern of homeobox gene expression in the early mouse embryo has a well-defined pattern along the main body axis. Some are

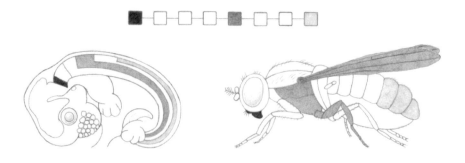

only expressed in the anterior region, others only further back. To the delight of all in the field there is also very good evidence that the genes expressed at different positions along the fly embryo are expressed in similar positions along the mouse embryo. The genes expressed at the anterior end of the fly are closely related to those at the anterior end of the mouse, for example. And in the developing mouse limb there is a pattern of homeobox gene expression that makes those genes strong candidates for being involved in the specification of positional information. Again, as in the fruit-fly, the order of the genes on the chromosome corresponds to the order of expression in the embryo.

Homeobox-containing genes provide an excellent example of how studies on animals apparently remote from human, such as flies, can provide fundamental clues to developmental mechanisms. Evidence is accumulating that even in primitive ancestral forms, homeobox-containing genes were involved in specifying the primary spatial patterns in developing embryos. They continue to do so.

WIRING THE BRAIN

THE WAY the billions of nerve cells, neurons, connect up gives us consciousness, language, emotions—all the things that make us human. While the nervous system is by far the most complex organ its development is based on essentially the same cell activities that are involved in the development of other organs. Because the nervous system is a network of neurons of varying shapes the central problem is how the neurons make the right connections. Some nerve cells in the brain have many extensions and so a single neuron can receive as many as 100 000 different inputs.

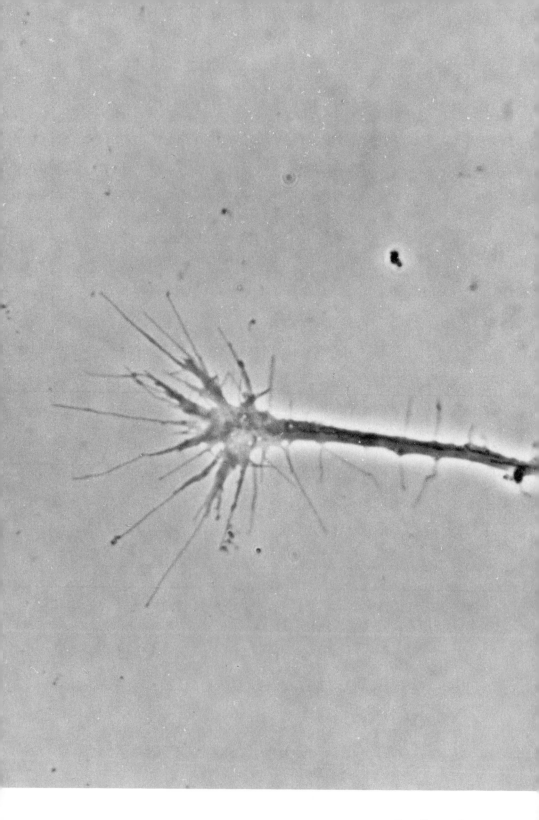

Growth cone of a neuron

MIGRATING AND CONNECTING

Migration and growth of neurons are linked to each other and are fundamental to the establishment of neural connections. Once the developing neuron has reached its proper location in the brain or spinal cord, as will be dealt with later, it begins to grow out first an axon which transmits signals and later dendrites, which receive signals from other cells. The first step in the formation of the major extension—the axon—is the development of a growth cone. In the manner of a hand with many long fingers this sends processes outwards to test the environment. These processes, like the filopodia of sea-urchin cells, (Chapter 2), are continually being projected outwards and being retracted, and their retraction pulls the growing extension forwards. The growth cone is a dynamic structure, its long processes testing the environment, often moving in the wrong direction, but eventually drawing the axon to the site where it will make the right connection. Recent evidence shows that the growth cone collapses when it makes contact with a 'wrong' region. There is also evidence that the growth cone will move in the direction of an increasing chemical gradient to which it is attracted.

Limbs of both embryonic grasshoppers and chicks provide excellent examples of pathway selection and guidance. Neurons that send sensory impulses from the tip of the grasshopper's limb to its central nervous system have their origin at the tip of the limb itself. The axons of these sensory nerves grow back along the embryonic limb to make connections with neurons in the central neural complex—zigzagging is a reasonable description of the path they take. The first axons are known as pioneers, for once they have marked out the path succeeding axons will crawl along them for guidance. These pioneer axons have to find their own way and they use specific cells *en route* as stepping-stones. A pioneer growth cone starts off by sending out long fine exploratory processes some of which are just long enough to reach the first stepping-stone. This provides a stable contact and the axon is drawn towards it by the extending and contracting filopodia. The sequence is now repeated as the growth cone seeks out the next

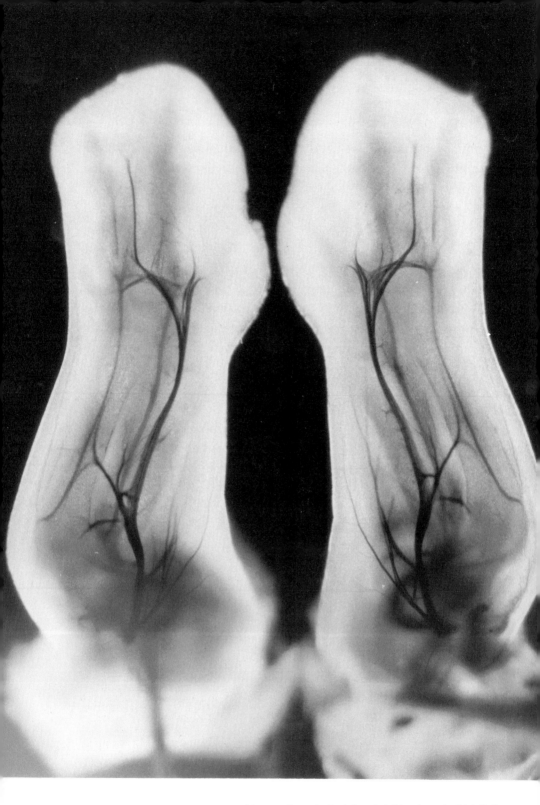

Nerve patterns in the left and right embryonic chick limbs

stepping-stone and so eventually the central nervous system is reached. Destroying a stepping-stone cell by laser microbeam has just the effect you would expect—the growth cone cannot find its way and

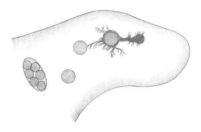

wanders aimlessly in many directions. Stepping-stones are not available to guide the growing neurons in other parts of the grasshopper and in those cases it seems that graded differences in the adhesiveness of the substratum provide guidance, the cells moving (again as in sea-urchins, Chapter 2) to more and more adhesive regions since these provide the most stable contacts. Mistaken steps are often taken but these are eventually corrected.

When axons from neurons in the spinal cord enter the chick limb bud they find themselves in a very large area filled with limb bud cells, for by this time the limb is already well developed. Unlike the grasshopper limb, there are no obvious stepping-stones and the neurons have a much greater choice of pathway. Yet the axons take well defined paths, first travelling together, and then branching off at just the right points. They make, it seems, relatively few mistakes.

For the motor neurons entering the chick limb that will connect up with muscles, the choices are rather like those faced by a driver on a multi-lane highway with numerous exits. How do the axons know which path to choose? One way to find out is to make the neurons enter the limb along the wrong pathways. Will they then be able to find their way? The experimental test is carried out by removing a small piece of the spinal cord opposite the limb bud before the neurons begin to grow out, and reversing its orientation. In this way neurons destined to connect with muscles near the anterior margin of the limb enter on a pathway taking them instead to muscles at the posterior part of the limb; and vice versa for neurons that go to posterior muscles. If the piece of neural tube reversed is not too large, so that the correct pathways for the neurons are not too far away, the neurons can still find their way to the correct muscles. If, however, the dislocation is too great, then the neurons will connect up with the wrong muscles. There is a limit to their ability to find the correct pathway and muscle, and they will connect up with the wrong muscle rather than no muscle at all. The selection of neuronal pathways and muscles is thought to reside in specific molecules along the pathway and on the muscles themselves. These have not yet been unequivocally identified.

FROM THE EYE TO THE BRAIN

Images falling on the retina of the eye are transmitted to the brain by the optic nerve which is made up of millions of axons from neurons in the retina and these axons make very specific patterns of connections in the brain. Rather like a television cable, the optic nerve carries the image to the brain and connects up with it so that the image is not distorted. This orderly connection was first discovered in 1943 by Roger Sperry the American neurobiologist, and another Nobel Prize winner, who studied the making of connections between the eye and brain of frogs. In frogs, the optic nerve connects up with that part of the brain known as the optic tectum.

Frogs like to eat flies. If a fly is seen by the frog slightly above its head, it quite naturally lifts its head and tries to catch it. What Sperry did was to cut the optic nerve between the frog's eye and its brain, and then rotated the eye through 180 degrees, completely inverting it. The severed optic nerve grew back from the retina to the optic tectum and a new set of neural connections were established. When the animals were tested by presenting them with a fly, they behaved as if their world had been turned upside down. A fly placed above the head caused a downward movement of the head, away from the fly. A fly placed below eye level resulted in the head moving up. No amount of practise could correct this inappropriate behaviour. The explanation of the frog's response was that the retinal cells had grown back to their *original* positions on the tectum, but now, because the eye had been inverted, an image on the retina from a fly above the head was transmitted to that part of the brain which would normally be stimulated when a fly was below the head. (By contrast, humans can quite quickly learn to live in an upside down world. Wearing glasses that invert the image takes a little getting used to, but subjects can even, eventually, ride a bicycle down narrow paths.)

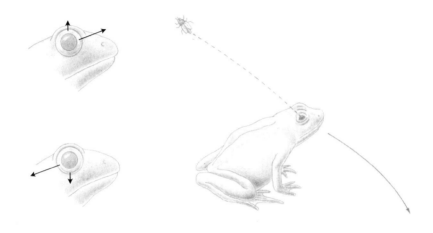

A map of the connections between the retina and the brain can be constructed by shining a narrow beam of light on to a particular part of the retina and then recording from the frog's brain the electrical activity due to the arrival of impulses from the optic nerve. The region where the optic nerve makes connections with the brain is the optic tectum and there is a point-by-point correspondence between the site of stimulation on the retina and the response on the tectum. For example, as the point of light is moved from the top of the retina to the bottom, so the electrical activity moves in a straight line from one side of the tectum to the other. There is thus a strict mapping of stimuli received by the retina on to the tectum.

If the experiment in which the frog's optic nerve is cut and the eye rotated is followed by mapping the connections of the optic nerve to the tectum, it confirms that the optic neurons did grow back to their original positions; just as the neurons which enter the chick limb are able to find their appropriate muscles. Sperry's plausible hypothesis proposes chemo-affinity, that is, the neurons are intrinsically different and carry chemical labels such as CAMs (cell adhesion molecules, Chapter 2) on their surface which enable them to match up with the appropriate neurons on the tectum. It assumes, in effect, a lock and key mechanism, thus enabling the neuron in the optic nerve to find, when it connects with the brain, perhaps just one other neuron—a little like recognizing a single face in a crowd.

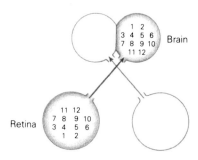

But this highly specific lock and key hypothesis runs into great difficulties, because neurons are, in a sense, more sociable, even promiscuous. One class of experiments looks at what happens when there is a disparity in size between the retina and tectum. What would one expect if half of the brain region were removed and the optic nerve cut? For a lock and key mechanism, the prediction would be that only those neurons which originally joined with the half of the tectum that remains would make connectins with it, and the remainder lacking their partners would make no connections. The regenerating neurons start off by connecting up in the expected pattern but, with time, this changes and *all* the neurons from the optic nerve make connections with just half a tectum. To do this the neurons that normally connect with the half tectum are 'squeezed' up into new positions. The result is a more or less normal mapping of retina on to tectum but with neurons making connections with quite new partners. This result illustrates that there is a lack of strict selectivity. It is rather like what we see in the limb when neurons are too far from their 'own' muscles and make instead connections with foreign muscles.

One should not completely discard chemo-affinity because of experiments similar to those just discussed. Rather, one should recognize that models other than lock and key type can be constructed. It may be more appropriate to think in terms of the neurons having molecules on their surface which affect their adhesiveness and that this adhesiveness is graded over the surface of the tectum. To this must be added something which seems to maintain neighbour relationships in the optic nerve: the neurons seem to try to keep the same neighbour relationship and this helps in ordering the map. The neurons then try to do two things at the same time: they try to make as strong adhesive connections as possible but also keep their relationship to each other. This compromise may lead to ordering of the connections. If this model were true then ordering of neural connections becomes closely related to the more familiar mechanisms of gradients and adhesiveness. It is satisfying to think that some aspects of the development of so complex an organ as the nervous system may begin to be understood in terms of the same cellular activities that are involved in the development of other parts of the embryo.

PATTERNING IN THE BRAIN

The brain and spinal cord develop from the neural tube that develops from a flat sheet of cells during neurulation (Chapter 2) after it has been induced during gastrulation (Chapter 3). That single sheet of cells gives rise to the complex layering of cells within the brain. Neurons are generated within a neural tube from cells that resemble stem cells in a manner not unlike that which we have seen in the skin (Chapter 6). Cells on the inner surface of the tube, for example, divide repeatedly and the cells they give rise to migrate away and become neurons. Since neurons themselves never divide they grow by putting out more branches or increasing their length. What kind of neuron develops is largely determined by when the neuron is born.

The cerebral cortex is that part of the brain concerned with higher mental functions. In humans, it contains two-thirds of the brain's neurons and is what sets humans apart from other species. Individual areas process sensory information, control motor activity, and are involved in language acquisition and spatial orientation. All these functions are carried out by a network of millions of neurons. The cortex is essentially layered, different layers containing special kinds of cell. For example, in the cortex some layers contain pyramid-shaped neurons, whereas other layers are quite different. Each layer has special connections both between the cells and with cells in other layers.

Patterning the cortex depends largely on the neuron's birthdays—for this seems to determine the developmental pathways of the neuron. A neuron has a definite birthday; it is the time of the last division before it becomes a neuron, for once differentiated as a neuron it will never divide again. The order of the neuron layers in the cortex corresponds to their birthdays. This comes about in the following way. The first cells to develop into neurons in the neural tube are

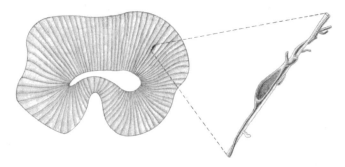

located near the inner surface of the tube. The cells divide and one of the daughters differentiates into a neuron. The cells migrate away from their site of origin, towards the other side of the sheet but only go a little way. The sets of neurons that are born later migrate even further and move past those that were born earlier, and so are even nearer the outer surface. Later born neurons migrate even further, so the youngest cells travel the greatest distance and end up outermost. In their migration the neurons are guided by special supporting cells— glia. These are very elongated cells that span all the way from the inner surface to the outer surface. Migrating cells cling to these glial cells and move along them, rather as one would climb a rope. The cells tend to remain in line with the parent stem cells thus forming a column of common descent.

Evidence for the importance of birthdays determining the developmental pathway, and so the character of the cells in the layers, comes from a study of a mutant mouse, *reeler*. As its name implies it walks in a most ungainly way, turning and tottering. The mutation in the *reeler* gene results in the incorrect migration of cells in the developing brain. The order of cells is almost reversed, with the old cells migrating and the later born cells remaining near the inner surface. In spite of their abnormal migration the cells differentiate not according to their new position but persist in their dependence on their birthday. Even though the oldest cells are in the wrong position they still develop

Reeler

pyramidal cells. The misplaced cells still try to make the connections they would have made if they had been in the correct position but the result, for the *reeler* mouse, is an abnormal gait.

The genes that control patterning in the brain are beginning to be identified. A number of homeobox genes (Chapter 7) are active in specific positions along the developing nervous system. It is very possible that they reflect positional information. Other genes involved are also coming to light. One, for example, which encodes for a protein associated with extracellular material, seems to be important for specifying the cerebellum. In a transgenic mouse lacking this gene no cerebellum develops. Transgenic mice, lacking specific genes, or which

contain new ones, provide a very powerful technique for finding out
how genes control development.

Cell death

Cell death is an almost constant companion of neuronal development.
In almost all parts of the developing nervous system too many
neurons are produced and the surplus ones die. For example, almost
half the motor neurons that initially enter the chick limb bud will be
dead within five days. If the limb bud is removed from the embryo
before the neurons enter, nearly all of the motor neurons in the spinal
cord die. In the complementary experiment an additional limb bud is
grafted on to the embryo and some of the motor neurons that would
have died are rescued. It seems that over-production followed by
selective cell death may have several advantages as a strategy for
constructing a neural network. It means, for example, that there is no
need to precisely match the number of neurons to the number of
neurons to which they will connect.

What then determines which neurons will survive and which will
die? In part the answer is that those that make successful connections
tend to survive. This in turn is due to chemical factors released at the
site where the connections are made.

Factors

The regions with which neurons make contact secrete chemicals which
are crucial for the survival of the neurons. These neurotrophic
substances, as they are called, were first discovered by a rather
circuitous route.

In 1948, a student of the American embryologist, Victor Hamburger,
found that if he implanted a mouse tumour into a chick embryo the
tumour was invaded by sensory neurons. The student did not pursue
the finding but both Hamburger, and another student from Italy, Rita
Levi-Montalcini, did. They showed that the effect was due to the
tumour releasing something which seemed to attract the sensory

neurons and which they called nerve growth factor. Levi-Montalcini then set out to purify the factor. She first managed to grow the sensory neurons in a culture dish. They put out, however, rather few processes. When she then added a small fragment of the tumour to the culture, there was a dramatic outgrowth of neurons. She had mimicked the behaviour of the neurons in the embryo in a much

simpler system. This provided her with a simple experimental system to which she could add extracts made from the tumour and see which caused outgrowth. She was then joined in the isolation project by an American, Stanley Cohen. At one stage they tried to find out whether the factor was a protein or a nucleic acid. So they treated their tumour extract with snake venom which contains an enzyme that breaks down nucleic acids; if the extract still worked then they could exclude nucleic acid. The extract did continue to work after treatment but they wisely tested the snake venom to see if, on its own, it had any effect—this is a routine control procedure. To their amazement the snake venom itself was a potent stimulator of neuron outgrowth and a rich source of nerve growth factor. Since the venom is made in the snake's salivary gland they took an imaginative leap and tested the salivary glands of male mice. There, they found a rich source of the factor which they were eventually able to purify. Nerve growth factor is a protein which acts at very low concentrations on sensory nerves and many other neurons. It is very likely one of the factors that prevents neurons dying when they reach their target. If it is removed from developing embryos then sensory and other neurons die. Nerve growth factors, and related factors, are of enormous importance in the development of the nervous system, and Levi-Montalcini and Cohen well deserved their Nobel Prize.

BEHAVIOUR AND DEVELOPMENT

In mammals, the first few months after birth is a time when experience modifies neuronal development. This is an important aspect of neuronal development which thus far I have ignored, concentrating on the embryo's developmental programme. Particularly in mammals like cats, monkeys, and man, those months after birth are a critical period in which visual experience of the world plays a vital role. The period is critical in the sense that during this time visual experience is essential. If the animal does not receive the right stimulus then irreversible and inappropriate connections may be made, with dire effects. An example is the so-called 'lazy eye' which can develop in children with a squint. Children who are born with a squint may fall into the habit of just using one eye with the result that the other eye does not receive well-focused images. If this situation is continued, and the squint not corrected, the unused eye becomes almost completely blind and the defect cannot be corrected. The defect is not in the eye but in the brain where the proper connections have not been made.

But even though experience can alter neural connections it is worth emphasizing that for many animals quite complex behaviour patterns are specified by the neural networks that are set up during development. Birds' ability to fly, the swimming of fish, and the walking of many animals all depend on neural networks that are laid down in the developing embryo. Even more complex programmes, like nest building and birdsong are, to a large extent, programmed during development. While it is true that much of the behaviour of higher mammals, particularly humans, is learned, it must be true that many of our higher mental functions are dependent on the networks generated during development. The psycholinguist, Noam Chomsky, has strongly emphasized that the ability to learn a language is a natural process that is programmed into the structure of the brain during development. Experience will modify such networks, but it would be hard not to believe that there must have been something special about the neuronal networks of the embryonic Newton and Mozart—a difference that might be traced back to neurons migrating and connecting.

SEX

SEX REPRESENTS a very interesting developmental problem. Males and females have very similar genes and yet are very different. How then is the developmental programme altered to produce the two sexes? Another problem relates to the germ cells themselves which have the remarkable property of eventually reproducing themselves.

Only the germ cells are potentially immortal, all other cells die. August Weismann, the great German biologist, understood this clearly when, at the end of the last century, he proposed the idea of the continuity of the germ plasm as distinct from the body cells. Whereas all body cells eventually die the germ cells that give rise to the next generation, in effect, have a continued life. During development all the cells which differentiate into body cells, that is nerve, skin, and so on, will eventually die with the organism, and only those cells which differentiate into germ cells may be passed on to the next generation.

The development of germ cells is just another example of cell differentiation. The similarity of the genetic constitution of germ cells to the other cells, and its constancy, is the basis of a fundamental biological principle: acquired characteristics are not inherited. The powerful muscles of the blacksmith's arms are not inherited by his children; a mother's knowledge of Russian is not inherited by her children; giraffes did not acquire long necks by their ancestors stretching their necks to the highest branches. Characteristics and attributes acquired by experience or learning are not passed on to the offspring. The reason is simple. There is no mechanism whereby the acquired character—strong arms, Russian—can be transferred to the germ cells and appropriately alter their genetic constitution. Even Darwin erred on this point, since he thought that the body cells contained particles—pangenes—which could be transferred to the germ cells and that this would allow for the inheritance of acquired characters. Disappointing though it may be, no such mechanism exists, and germ cells only pass on the genetic information they start with, together with any random mutations that may have arisen during the cells' lifetime.

Sperm are machines for delivering the male DNA to the egg. They are highly motile and contain little more than the male nucleus and a means for locomotion, usually a long undulating tail or flagellum. Its sole purpose is to fuse with an egg and since millions are produced there is a fierce competition and the vast majority must fail in their short life's work—only a very few of the many millions ever fertilize an egg.

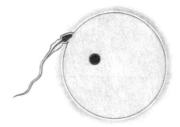

Fertilization of an egg in a test-tube is really a very simple procedure. With sea-urchins, one simply adds a drop of sperm from a male urchin, diluted with sea water, into a dish containing the egg. The sperm, millions of them, swim at random and when they encounter an egg throw out a rod-like structure that helps them first fuse with and then enter the egg. The egg for its part now sends a signal around its surface to prevent more than one sperm from entering. The fertilization of human and mouse eggs in a 'test-tube'—really a glass dish—is essentially the same. However, some important advances were required to make sure that the eggs and sperm were in the right sort of chemical solution to make them function properly.

The most important feature of fertilization is not that it causes development to start—there are lots of ways of getting eggs to do that. Almost any 'kick' will set the egg off—acid, alkali, or electric shock. The essential feature is the introduction of the nucleus of the sperm which carries the genetic information from the male. This nucleus fuses with the nucleus of the egg and together they provide the genetic programme for development.

MALE AND FEMALE

From the moment the sperm has fertilized the egg the genetic sex of the animal is determined. The sex is determined by the chromosomal composition of the male nucleus and not, as Erasmus Darwin, Charles Darwin's grandfather, used to think, by the thoughts of the father at the time of conception. It was a widely held belief in the eighteenth century that the character of the child was entirely determined by the father and that the mother provided little more than a nutrient medium. Even at the end of the last century, heat (if not passion) and nutrition were thought to be major factors determining the sex of the child; factors favouring the storage of energy were thought to promote female offspring, whereas energy utilization favoured male offspring. Environmental theories collapsed with the discovery of the sex chromosomes.

There is a difference between the genetic sex of a mammal and its sexual character. The genetic sex is determined by the X and Y chromosomes but the sexual characters, such as breasts and genitals, are determined by hormones made by the testis and ovaries. What the genetic sex does is to determine whether or not a testis develops and so whether or not the male hormone testosterone is made in the embryo.

In mammals, including humans, there are two sex chromosomes, the X and the Y. The Y chromosome is the key factor determining the sex of the embryo. The egg always carries an X chromosome and the sperm carries either an X or a Y. If the fertilized egg has an XX complement of sex chromosomes it will develop as a female, if it is XY it will develop as a male. So every cell in the female embryo will then have an XX complement of chromosomes and every cell in the male,

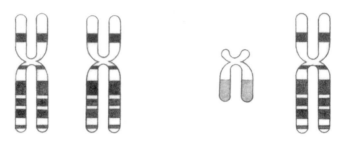

XY. Yet whether an XX or XY is present, has, for the vast majority of the cells in the body, absolutely no influence on whether the animal develops into a female or male. The only cells that are affected by the sex chromosomes are the germ cells and those in the tissue that will give rise to an ovary or a testis. All the main sexual characteristics that make males and females different arise from the effect of the hormonal secretions of the testis. As always, the hormone acts by changing gene expression.

The remarkable dependence of differentiation of sexual characters on the ovary and testis was first demonstrated by preventing either a testis or an ovary from developing in the rabbit embryo. While they were still being carried in the womb, the tissue that would form testis or ovary was removed and it was found that all the offspring had female characters irrespective of whether they were XY or XX but, of course, lacked ovaries. The conclusion was that all embryos would develop into females unless a testis developed. Testosterone, the male hormone made in the testis, was then shown to be the controlling factor which directs the embryo away from a female pathway into becoming a male. Secretion of testosterone causes the tissues of the body, irrespective of whether they are XX or XY, to take on male characteristics. Embryos are thus basically female but are converted into male form by the hormone testosterone. Since only XY embryos develop a testis and so produce testosterone, the problem then reduces to why XY embryos make a testis?

The testis and ovary develop from the same tissue and at an early stage there are no indications which will develop. Then there is a divergence: in XX embryos the tissue develops into an ovary, whereas in XY embryos it develops into a testis. The development of a testis is absolutely dependent on a gene on the Y chromosome. Near the end of the short arm of the Y chromosome sits the crucial male-determining gene, somewhat isolated; for there are few other genes on the Y chromosome. This testis-determining gene is active in the early embryo only in the region of the developing testis. If the gene is deleted then even XY embryos develop as females. But if the gene is translocated to the X chromosome, as can happen in mice, then XX embryos now develop as males. In both these examples there is a sex reversal with respect to the gross chromosomal composition. Given the importance of the testis-determining gene we are still left with the problem of how this gene's product leads to the development of a testis. One can only assume that there is a cascade of gene activations and cell interactions, which remain to be worked out.

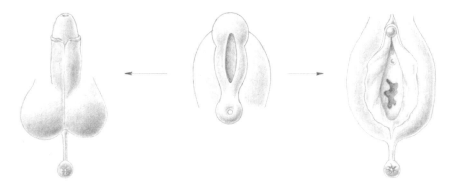

While the Y chromosome determines the genetic sex of mammals and whether or not a testis develops, all the so-called secondary sex characters, like the genitals and breasts, are determined by hormones. In particular the testosterone produced by the embryonic testis causes male sexual characters to develop. After six weeks of development the genitals of a human embryo are made up of a series of rather dull swellings and there is no difference between males and females. But the primary sex has already been determined, and now the secondary characters come under hormonal control. In the absence of testosterone, which is made by the testis, the swellings develop into the female genital organs—labia and clitoris. But in the presence of testosterone, the same tissues develop into a scrotum to enclose the testis and a penis. Similarly, in the absence of testosterone, breasts develop in females and testosterone blocks their growth in males. Again, facial and body hair develop when testosterone is present, and so too does baldness. Embryos are basically females unless exposed to male hormone.

A mixture of female and male characteristics occurs if there is a fault either in hormonal secretion or in cellular response to the hormone. For example, there is an inherited human abnormality known as testicular feminization in which the genetic constitution is XY and a testis is present, but all the secondary sexual characters are female. While testosterone is produced by the testis the cells lack the receptor to respond to the hormone; they thus develop as if no testosterone were present. The converse case is the result of abnormal hormone secretion during development of a female embryo resulting in a genetic female resembling a male but still having an ovary. Such people are not hermaphrodites, since they have either an ovary or a testis, but not both. They are unlike the nematode worm and some fish which are true hermaphrodites since they can have both ovary and testis.

In the fruit-fly *Drosophila*, while sex is determined by chromosomal composition, the sexual structures are not a result of hormone action. Rather, each tissue develops independently as male or female, depending on its chromosomes. Nor is sex determination based on a Y-like chromosome but on the number of X-like chromosomes. Two X chromosomes give a female and just one results in a male. Since the development of sexual characters is autonomous and controlled by the chromosomal composition it is possible, unlike mammals, to get animals in which parts are male and other parts female. If an X chromosome is lost from one of the nuclei of an early female embryo, all the descendants of that cell will be male while all other cells in the embryo will develop female characters.

Environmental factors, rather than genetic constitution, determine the sex of turtles and some alligators. For them, temperature is the controlling factor. In general, eggs incubated at low temperatures produce one sex only whereas a higher temperature results in the other sex. (The eggs are laid together in heaps, and complex arguments have been put forward to account for the supposed evolutionary advantage of temperature control of sex.) Temperature may activate (or inactivate) a testis-determining gene similar to that found in mammals and so make the developmental pathway rather similar. There is a speculation that dinosaurs, the ancestors of turtles and alligators, may have used such a mechanism. A slight change in the earth's temperature may have created conditions where just one sex was produced. This could, in principle, provide an explanation for their mysterious disappearance.

MOTHERS NEED FATHERS

Are fathers really necessary? Could the female nucleus support development on its own as assumed in virgin birth? For some lower vertebrates there is no doubt that this is possible. Frog eggs can be artificially activated and will develop into reasonably normal frogs even though each cell has only half the normal number of chromosomes. Mammals, on the other hand, behave differently and eggs, even when activated to start developing, fail to do so. This is not due to just some minor defect, but represents a fundamental difference between the female and male contributions to development.

Recent studies have shown that the genetic information in the male nucleus brought in by the sperm and that in the nucleus of the egg controls different aspects of the developmental programme. The egg's genes contribute primarily to the development of the embryo proper, while the sperm's genes control the development of extra-embryonic structures, such as the placenta. Both nuclei are required for normal development.

Since both egg and sperm nuclei have only half the normal number of chromosomes, a clever experiment was designed to provide the normal number of chromosomes in all male or all female embryos. This was done by removing the male nucleus from a fertilized mouse egg before it fused with the female nucleus, and replacing it with a female nucleus from another egg. The two female nuclei fused quite happily, restoring the normal number of chromosomes, and development proceeded quite normally. But only for a while. When the embryo was quite well advanced, with head and somites, development ceased, because the placenta was poorly developed. If, on the other hand, two male nuclei were placed in an egg whose own nucleus was removed, development again started normally but ceased at a much earlier stage, the embryo itself being abnormal.

These results provide evidence that there is a difference in the behaviour of genes depending on whether they were contributed by the mother or the father, so that the genes of the female made a special contribution to the embryo proper, while the male genes were programmed to control placental development. Female and male genes, even though they have similar genetic information, have had a different pattern of gene activities imprinted on them. So the reason development of the all-female embryos fails is because the placenta and other supporting structures do not develop; and the all-male embryo fails because it cannot make a proper body. Mammalian embryos require contributions from both male and female: virgin births are not possible.

SEX AND BEHAVIOUR

Behaviour of many animals is sex-specific. Male rats mount female rats; male canaries sing eloquent mating songs. Testosterone affects not only obvious sexual characters, like the genitals and breasts, but also alters brain development, and thus these behaviour patterns. In songbirds a close correlation has been demonstrated between male song and testosterone levels. If the testosterone level is too low chaffinches will not sing but if testosterone is injected into males, the mating song will be sung even outside of the breeding season. Again, in rats the sexual behaviour is controlled by early exposure to testosterone. Newborn female rats treated with testosterone grow up to behave like males and try to mount other females, and do not show the characteristic behaviour of female rats. This changed behaviour may be traced to the effect of the testosterone on specific spinal neurons that may be involved in sexual behaviour. Female rats lose 70 per cent of the neurons in that spinal region whereas only 25 per cent of the neurons are lost in males.

There is both a temptation and a danger in extrapolating from birds and rats to humans. Humans do not have highly stereotyped mating calls or sexual behaviour patterns. Nevertheless, it is hard to believe that hormonal influences have no influence on the brain, leading to males and females being different in their behaviour. Environmental influences may be very powerful, but to deny differences between the brains of females and males may be to miss the importance of a fundamental developmental process.

GROWING

GROWTH IS an integral part of the developmental programme. During early development there is little growth in the sense of increase in size of the embryo, and, as we have seen, the changes in form of the embryo are largely due to patterns of localized contractions and cell movements. However, at later stages of development growth plays a key role and is intimately involved in moulding the form of the embryo as, for example, the face. The growth of the embryo is dependent on nutrients and in the case of the chick and the frog this is provided in the egg as yolk. Once the basic form of the embryo is established blood vessels carry nutrients from the yolky regions to all parts of the embryo and so permit growth to occur. In some small embryos such as the sea-urchin and in insects a circulation of nutrients is not necessary, since the yolk is distributed amongst all the cells. For mammals a special circulation brings in nutrients via the placenta.

The programme for growth is specified at an early stage even though this programme may take years to complete. A key feature as regards the limb (Chapter 4), and many other organs, is that the basic pattern and structure is laid down on a very small scale. All the main features of the human limb, for example, are laid down when it is no more than a few millimetres long, some four weeks after conception.

Similarly, all the main features of the body are already present in miniature by five weeks of age, when the fetus is less than an inch long. It is growth that expands this miniature to its adult size.

In addition to this expansion the form of many parts of the body is determined by how much they grow in relation to one another. Subtle differences in growth of parts of the embryonic face affect its final form. Even during the fetal period, different parts grow at different rates: for example, the relative size of the head of the human embryo compared with the rest of the body becomes continually less with development.

THE PROGRAMMING AND CO-ORDINATION OF GROWTH

We can take the growth of the limb as our model for growth of an organ. The programme for growth of the vertebrate limb and its individual elements is laid down very early in development. Thus the characteristic growth rate in the different bones of arms and legs is specified when these elements are themselves laid down. The growth of the early cartilaginous elements seems to be autonomous in the sense that it is independent of the age of the rest of the embryo. If a chick limb bud is grafted from a young embryo to an older one, or vice versa, the growth characteristics of the grafted limb are uninfluenced by the host. The grafted limb grows as if it were still in its original embryo.

The autonomy of the growth programme is dramatically illustrated by a 1924 experiment done by Ross Harrison, one of the pioneers of American experimental embryology. There are two species of the axolotl, an amphibian that retains juvenile features, that differ very appreciably in size. Harrison grafted a limb bud from the embryo of the larger species in place of the limb bud of the embryo of the smaller. Initially, the grafted bud grew more slowly than the host's buds, but it then increased its growth rate and eventually grew to its normal size: the small axolotl had one very large leg.

But there is dependence as well as autonomy. The growth of the bones is autonomous, provided the right hormones are present, but the growth of the muscles and tendons are dependent on bone growth. It is this dependence that ensures that bones, muscles, and tendons maintain the proper relationship during growth. The co-ordination of the growth is achieved mainly by mechanical means; the growing bones pull on the muscles and tendons making them grow. It seems that the stimulus to the growth in length of muscles and tendons is just tension—pull. If for example, the growth of a bone is delayed, so too will be that of the associated muscles and tendons. In this way, the lengths of muscles, tendons, and bones are nicely adjusted to each other.

There is, however, nothing to co-ordinate the growth of the two arms and two legs. Their growth is quite independent of each other; there is nothing to tell one arm how fast the other is growing, there is no information passing between them. So it is remarkable that the limbs, that are laid down as tiny miniatures in the embryo grow independently for the next 15 years and end up almost exactly the same length. It is like setting off two trains and finding that 15 years later they have travelled the same distance. Cells are clearly very reliable at carrying out very long programmes of growth.

Cell multiplication

The programming of cell multiplication is the main feature of growth, but by no means the only one. Growth can also take place by the cells getting bigger rather than multiplying, neurons providing a clear example. Another major mode of growth is the laying down of materials outside the cell, thus increasing the extracellular space. The growth of finger and toe nails is an example of the laying down of extracellular material; and so is the matrix between cartilage cells. Even so, cell multiplication is the real key to growth.

The decision whether to multiply or not is fundamental to the life of the cell. In adult mammals different cell types take quite different decisions. Some cells, like stem cells in the skin and those lining the gut, multiply continuously to replace lost cells; others, like those in the liver, do so rarely but can be stimulated to multiply when injured; while yet others, like mature nerve and muscle cells, never multiply. In each case the decision whether to multiply or not is tightly controlled: loss of this control leads to cancer (Chapter 11).

The control of cell growth and multiplication in the embryo is poorly understood. It is not known how a programme for cell multiplication is specified in the cells. However, there are factors which stimulate growth and these may be produced locally or, like hormones, be carried throughout the embryo or body by the circulation. Growth hormone, as will be seen, is of particular importance. But even in the presence of such growth-stimulating substances, cells can respond in quite different ways. How the cells respond is part of their developmental programme.

GROWING TALLER

Human growth from a small baby to an adult involves all the organs but is most obvious in the trunk and legs. The growth in length of the limbs is entirely due to the growth of the bones like the femur and tibia in the leg. Since bones are mechanically rigid they cannot directly increase in length and so there are special growth plates near the ends of the bone which provide for the increase. These growth plates are not bony but are made of cartilage like the original cartilage model of

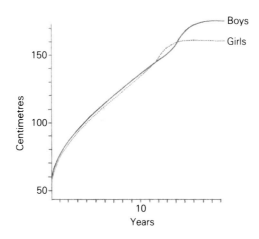

the bones that are first laid down. Unlike bone, cartilage is more flexible and can expand when cells divide or get bigger. This is the brilliant biological solution to increasing the length of bones while maintaining mechanical strength.

The growth plates are just a few millimetres thick and are located near the ends of all the long bones, such as the humerus and femur. In the growth plate the cartilage cells are arranged in columns, with stem cells at one end. Overall, the growth plate is similar to the lining of the gut (Chapter 6), but instead of the cells being lost they are replaced by bone.The cells continue to multiply as they are pushed towards the other end where they stop dividing, and are replaced by bone, which is laid down by other cells. The thickness of the growth plate remains constant during growth, and as the cells at the end are replaced by

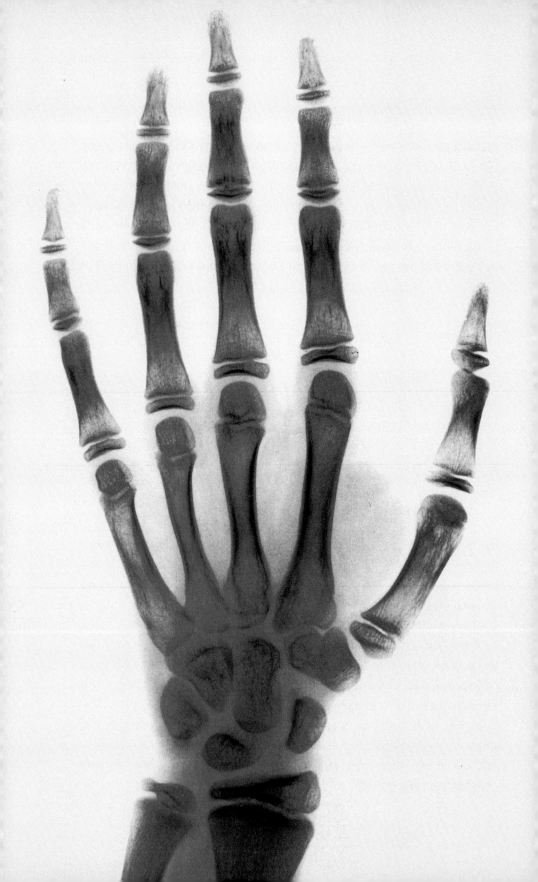

bone, the whole bone gets longer. Each growth plate has its own particular growth characteristics. With increasing age, the growth plates disappear and are completely replaced by bone, so that growth now stops completely. The loss of growth plates occurs in a very specific order and can be used as a measure of a child's physiological age.

Human growth rate in the first few months after birth will never be surpassed. As we continue to grow the rate of growth, that is the increase in height per unit time, gets less and less, until, at adolescence, it increases dramatically. There is a rapid increase in growth at adolescence—the adolescent growth spurt—which occurs in girls at about 14 years old and in boys a year or two later. In fact there

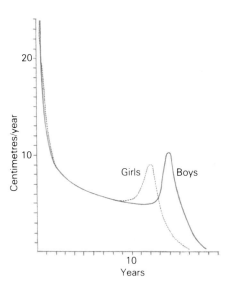

is a great deal of variability in the time when the growth spurt occurs: this is linked to when puberty takes place. The growth spurt is short-lived and is followed by a complete cessation of growth. Cessation of growth is directly linked to the disappearance of growth plates. Once they are replaced by bone they never reappear and no further growth can occur.

The growth of hundreds of normal children has been plotted, and this has provided a set of standard curves which enables doctors to

Growth plates in a child's hand

decide whether a child's growth is proceeding normally and can also give an early indication of how tall the child will eventually be. For a girl who wishes to become a ballet dancer, where acceptance is restricted to a fixed height range, this can be helpful if often disappointing. More importantly, comparison with these standard curves can quickly show any abnormality in growth. The most serious and common of these are those disorders which lead to restricted stature—dwarfism of different kinds.

The growth plate is dependent on growth hormone in the blood that is produced by the pituitary in the base of the brain. In the absence of a sufficient amount of circulating growth hormone, the growth plates grow more slowly or may even cease to grow at all. Other organs also grow more slowly so that the child is normally proportioned, but is disproportionately small for its age, and unless the growth hormone levels are restored the child will be very short. Fortunately, this is now possible, and daily injections of growth hormone can restore normal growth. Growth retardation can also result from thyroid deficiency, starvation, and psychological trauma. All these can be corrected and result in the child reaching its normal height. If, on the other hand, growth plates continue to function longer than normal, heights of over eight feet can be attained.

A quite different cause of restricted growth is not related to hormones but is caused by the growth plate being disorganized, the general condition being termed achondroplasia. Only the plates are affected, so only the limbs and trunk suffer from growth retardation, leading to an adult who is short in stature and whose head seems disproportionately large. Thus far no treatment is available satis-factorily to treat defective growth plates.

What then determines our height, and why, for example, are pygmies short? Other growth factors, whose synthesis is stimulated by growth hormone, play a key role in growth, and they increase during the growth spurt. Although pygmies have normal levels of growth hormone, the level of one of these growth factors is much reduced. The growth spurt is brought about by an increase in growth hormone which in turn is brought about by an increased concentration of the sex hormones. Cessation of growth is mainly due to disappearance of the growth plates under the continued influence of these hormones.

So a major feature of growth is the response to the hormone and the characteristics of the growth plates, and these are specified early in development.

RELATIVE GROWTH

Different parts of an organism may grow at different rates. The head of a baby is disproportionately large and it grows more slowly than, for example, the limbs. This difference in relative growth rates can have profound effects on body form. A striking example is shown by the claws of the male fiddler crab. When the male crab is still small the two

claws are of equal length. As the crab grows, one of the claws, the large crushing claw, grows much faster than the other and ends up weighing about four times that of the other.

Relative growth has, as will be seen, important implications for evolutionary change. From the point of view of growth control it emphasizes how different parts of an organism have different growth programmes, which are laid down early in development. Differences in local growth programmes can markedly alter the form of a particular structure, such as the skull. There is a big change in the form of the skull of a baboon as it grows up; when it is newborn the face is

relatively flat, but the jaw and nasal portion grow much faster than other parts giving the adult baboon its characteristic protruding face, the adult male skull being much larger than that of the female.

The problem then is how these differences in growth could be programmed. While the answer is not known it seems likely that regulatory genes and the positional values of the cells will turn out to be involved (Chapter 3). For, in order for parts of the body to grow to different extents, the parts themselves must be different, and positional values could provide just that difference. For example, the

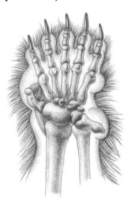

different positional values of the digits (Chapter 4) could be used to programme different patterns of growth, giving, for example, the elongated digits of the bat wing. A remarkable case of relative growth giving an extra digit is provided by the panda's 'thumb'. Pandas strip bamboo, their stable food, by passing the stalks between what looks like a thumb and the other five fingers. The 'thumb' however, is a digit not at all like the others. It is an enlargement of one of the bones in the wrist that is so elongated that it has all the appearance of a typical thumb. Enlargement and increased growth of a single bone has thus generated a quite new structure. This can only have been possible if that bone had a positional identity that allowed its programme of growth to be uniquely altered.

CELL MULTIPLICATION AND CANCER

Cancer can be thought of as a breakdown in the developmental controls. Cancer cells have thrown off the normal controls of cell growth and multiplication and in many cases this is linked to the breakdown of the normal processes of cell differentiation. Furthermore, the most malign characteristic of cancer cells results from the loss of control of cell migration, and this enables cancer cells to spread from the site where the tumour first formed to distant and numerous sites. Cancer cells are characterized by these two properties—their multiplication is not subject to normal controls and they invade and colonize other tissues, which is known as metastasis.

THE CELL CYCLE

Many different kinds of normal cells will grow in culture. Cells can be taken from the early chick embryo or from beneath human skin connective tissue, placed in a dish, and if they are provided with the right culture medium they will attach to the dish and begin to grow and multiply. Cells first grow, double in content and size, and then divide to give two cells; this typically takes about 12 to 24 hours. Division of the cell takes about one hour and involves the distribution of one set of chromosomes to each daughter cell. Then the daughter cells can begin to grow again, and repeat the cycle of growth and division. From the

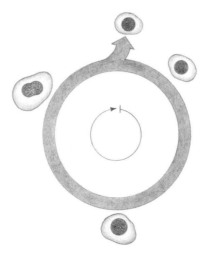

point of view of control of cell multiplication, the key decision is whether or not to start growing, and this is taken near the beginning of the cell cycle, known as the restriction point.

For the cells to survive for a while in culture they need a balanced salt solution and an energy source, such as glucose. This solution alone, however, will not permit cell multiplication: special growth factors are essential. These play a fundamental role in controlling cell growth. A typical growth factor is insulin and insulin-like molecules. All growth factors are proteins and there are receptors on the cell surface to which they attach and so stimulate growth. Different cells have different requirements, but in general, increasing the concentration of the growth factor increases the number of cells multiplying and the density at which multiplication will stop.

When the cells are placed in culture they multiply until they cover the bottom of the dish as a continuous monolayer. Then they stop multiplying and all the cells sit quietly at the restriction point. If the cells are now removed from the dish and plated out at lower density in further dishes, they will enter the cycle again and continue multiplying. Cancer cells do not grow and multiply faster than normal cells. The fundamental difference is that they continue to grow under conditions in which normal cells would stop multiplying. Thus, cancer cells will grow to much higher densities than normal cells when placed in culture. Most importantly, cancer cells have a much lower requirement for growth factors. The cells have changed such that they behave as if the growth factors were present even when they are not. The reason for this behaviour could be that the cells continually produce their own growth factor and are thus self-stimulating. Another cause is that the surface receptors for growth factors, or part of the pathway from the surface, are so altered as to be permanently activated and are continually in the switched-on state, and so the cell always behaves as if growth factor were present. Cancer cells thus escape normal controls of growth.

ORIGINS

Like developing embryos, tumours have their own developmental programme; but unlike embryonic development these abnormal programmes are highly variable. Tumours usually start as a mild disorder of cell behaviour that slowly develops into a full-blown cancer. The progression from the earliest event to the clinically recognizable tumour can take many years—it is a developmental process. Good evidence for the slow progression from initiation to tumour comes from the increased incidence of leukaemias in those who were exposed to radiation when the atom bomb was dropped on Hiroshima. Radiation damages the DNA and is known to be able to induce cancer. There was no significant increase in leukaemias until about five years after the bomb. Again, workers who have been exposed to carcinogens, chemicals that cause cancer, may not develop tumours until 10 or even 20 years after exposure. Part of the reason for this long period is that the cells have to undergo a number of changes before they become truly cancerous and malignant.

Normal cells do not change into cancer cells in just one step, but, like differentiating cells, they must go through a number of stages.

Cancer cells have their own developmental pathway. One of these stages in this path seems to be the acquisition of immortality, that is the ability to undergo unlimited multiplication. Normal cells have a limited capacity for multiplication (Chapter 12). Other stages enable the cells to escape from the normal controls of cell multiplication and differentiation and to migrate to and colonize other tissues. Each of these stages probably requires a specific change in the genetic constitution of the cell, a mutation. Thus the development of cancer can be thought of as a series of 'hits' on the cell's DNA, each one carrying the cell closer to true malignancy. Because each 'hit' is in itself a rare event, the chances of a single cell acquiring all of them by chance are rare but with the passage of time it is more likely that all the necessary events will happen in the same cell. It is for these reasons that cancer is largely a disease of old age. There are, however, rare cancers that affect children. These arise because the child has already inherited defective genes so when it is born the number of additional changes required for a tumour to develop is reduced.

The developmental nature of cancer is confirmed by the observation that most tumours arise from just one single cell. That single cell can generate the millions of cells that form the tumour. Evidence for this clonal origin of tumours comes from the chimaeric make-up of the female (Chapter 6). Women are made up of a random mixture of two kinds of cells which differ in which X chromosome has been inactivated, the maternal or the paternal. When certain tumours occur in a woman the cells in them are of just one kind, they either all have the maternal X inactivated or all have the paternal X inactivated. This strongly suggests that the tumour arose from one single cell. Similarly, in cancer of blood cells—a type of leukaemia—all the cells have a clearly identifiable change in one of the chromosomes. It is very unlikely that this change, which is a very rare event, could have occurred in more than one cell at the same time.

Cancer cells can also arise from disturbing normal development. If the early mouse embryo is grown in culture instead of being returned to the uterus, then the organization of the embryo breaks down. Individual cells separate out and multiply. The cells show no sign of differentiating and seem capable of multiplying indefinitely. These cells have become malignant cancer cells. If even a few cells are injected into a mouse a tumour will develop which will contain a variety of differentiated cell types. It is remarkable that freeing the cells of the normal mouse embryo from their normal environment should produce a dramatic change. Even more remarkable, the change is reversible. If cells are restored to their normal environment in the early mouse embryo then they will respond by contributing in a normal manner to the development of the embryo. This is a special case of a cancerous state being reversed by returning the cells to the right environment.

An evolutionary type of process plays an essential role in generation of a tumour. When normal cells, like liver, multiply, the daughter cells are like the parent cell. With cancer cells there is much more variation, daughter cells differing from the parent. Because of this variation in the tumour cells there is selection, regrettably, for those cells that multiply best and are most fitted to resist the body's defences. That the cancer cells will eventually kill the host in which they are reproducing, and so they are in a sense committing suicide, is irrelevant. The more a potentially malignant cell multiplies, the greater the chance that one of its offspring will take the next step towards a more cancerous state.

Cells that are normally multiplying are much more likely to give rise to cancer than other cells. By far the great number of cancers are carcinomas, that is they are derived from cells that normally give rise to the linings of the body, the skin, and the layer of cells on the surface of the gut. These cells are constantly multiplying to replace the cells that slough off (Chapter 6). The precursors of white blood cells, too, are constantly multiplying, and thus leukaemias—cancer of white cells—are quite common. In the normal replacement of skin, gut, and blood, the cells multiply and differentiate, although the final products are differentiated cells like skin cells and blood cells which do not themselves divide. So a major feature of the development of cancer is the failure of the cells to undergo normal development and differentiate into cells that do not multiply. For example, in several forms of leukaemia, it is as if the cells were blocked along the pathway to their mature state and so they continue to multiply without restriction. Multiplying to generate mature blood cells is normally a short period in their life history; now it is their whole life, since their maturation has been blocked. If substances could be found which could force the cells to complete differentiation then multiplication would stop.

Malignant cells that metastasize must migrate across all sorts of boundaries that present barriers to most normal cells. Melanoma cells, which are cancer cells derived from pigment cells in the skin, are highly invasive. They not only lose contact with their neighbours, which facilitates migration, but are able to get in and out of blood vessels. The normal controls of cell movement are lost. The differences in invasiveness in a population of melanoma cells has been demonstrated by isolating individual cells, growing up in a population from each cell separately, and then testing their ability to metastasize. This ability varies widely from the highly invasive to the almost benign. But if individual cells of a mildly invasive tumour are isolated and used to generate new populations, among these new populations highly invasive populations will be found. Cancer cells do not breed true, and new variants are constantly arising.

TUMOURS AND BLOOD VESSELS

Judah Folkman has often made an offer to his research students at Harvard University of a holiday for two, for two weeks, at the best hotel in Miami, to anyone who could culture a tumour that grew to more than two millimetres in diameter. Like cells, tumours can be grown in a culture medium in a dish. But these cultured tumours are always small, minute in fact, quite different from tumours growing in the body which can be many centimetres in diameter. Folkman's prize is quite safe, for tumours need a blood supply in order to grow, and outside the body, in a culture dish, they remain small. Folkman's point is that there is an intimate relationship between tumours and the blood vessels that supply it. The tumours actually attract vessels to grow towards them. If blood vessels do not supply the tumour with oxygen and nutrients, then the tumour either remains tiny or it will die.

Tumours release substances that attract blood vessels towards them. The vessels send out cells which move up a gradient in the attracting substance and new vessels are drawn to the site of the tumour. Once the tumor receives a blood supply its growth is explosive. Tumours can grow to 16 000 times their original volume within a few weeks. It may be that there are many latent very small tumours in the body that continue to remain unnoticed until they are vascularized. The development of local blood vessels could have fatal consequences. So the inhibition of the development of blood vessels near the tumour could lead to a new approach to controlling cancer.

GENES AND CANCER

Genes control the development of cancer but unlike embryonic development where they provide a programme for the orderly emergence of cell types, the genetic control of cancer involves random mutations of genes. Cancer develops because genes are altered. Only a small class of genes are thought to be involved. These genes—called oncogenes—are normal genes whose activity is altered during the development of the cancer. Oncogenes are normally involved in some way in the control of cell proliferation and differentiation and an alteration in their activity can lead to a breakdown of this control. For example, some oncogenes code for growth factors and in their altered state, the gene is switched on all the time so that the cell is continuously secreting a factor that stimulates its own multiplication. Other oncogenes code for proteins involved elsewhere in the network that controls cell multiplication: the result of an alteration in the gene is, again, continued and uncontrolled cell multiplication. Yet other genes control the activity of oncogenes, and mutations in these genes can also lead to the cancerous stage. This is because the circuitry for promoting multiplication is permanently switched on, and the pathway to normal differentiation blocked.

AGEING

DEVELOPMENT DOES not stop at birth. The developmental programme continues in some animals for many years, illustrated, for example, by the programme for growth. With time, organisms age and it is far from clear to what extent this is part of the developmental programme. By ageing I mean senescence and that the likelihood of death increases with age and that this is due to deterioration in some aspects of the organism's function. In this sense ageing is not only quite different from growth but there is evidence that continued growth even tends to prevent ageing. Ageing might be thought of as becoming a dominant feature only when the developmental programme ends. Consider a mouse and an elephant. If their eggs are fertilized on the same day then, at the time when the elephant is born, the mouse is already old. Elephants are born after 15 months' gestation, while a mouse 15 months old has only a short while to live and is ageing. Since both the cells and developmental mechanisms in elephants and mice are very similar we need to understand ageing in this context. Why do some animals age early and only live a short time while others live for a long time? If embryonic tissue does not show signs of ageing and there is no reason to believe that the tissues of the newborn elephant have aged—there are no signs of deterioration or loss of function—then it may not be unreasonable to think of ageing as being somehow related to the loss of the embryonic state.

The age to which animals live is under genetic control, and this age is related to reproduction. Humans can live to a very old age compared to mice, but then mice reproduce when they are very young and do not survive in the wild much beyond a year—they are usually eaten by predators. It is only in the laboratory, where they are protected, that mice live for several years. From an evolutionary point of view this makes sense since there is no point in designing an animal to survive for 10 or 20 years if none of the animals ever reach anything like that age. The most sensible strategy is to make sure that an animal is in peak condition when it reproduces and cares for its young. Every effort should be made to prevent deterioration of function before that time. So it is really no surprise that the elephant shows no sign of ageing when it is born, or that the mouse of similar chronological age is in every sense aged and shows numerous signs of deterioration.

Another way of looking at the same problem is to think of the animal as a machine designed to deliver the germ cells that are required for reproduction. An investment in construction and repair must be as economical as possible, and this leads to a design specification for a machine that delivers and then, its function fulfilled, just falls apart. So the investment is put into making a delivery machine, the body or soma, with only sufficient repair processes to ensure that it will remain in good condition until reproduction is over. On the other hand, it is worth making an investment that ensures that the germ cells do not age. This approach to ageing has been called the disposable soma theory.

In evolution there is thus selection only against ageing processes and diseases that influence reproduction. Once reproduction including care of the young is over there is no selective advantage in surviving. It is for this reason that so many diseases beset old age in our society. Even a few hundred years ago the majority of the population died well before they became old and there is little, if any, selective advantage for the offspring in keeping the parent alive after child-bearing and child-caring age. There has thus never been selection against cancer or cardio-vascular disease for people in middle to old age. Advances in medicine and improvements in public health have increased the number of people achieving middle to old age and the diseases that then occur are a natural result. In spite of the increase in the number of people getting older, the maximum age to which people live, around 110, seems totally unaffected. It is an age barrier whose nature we just do not understand.

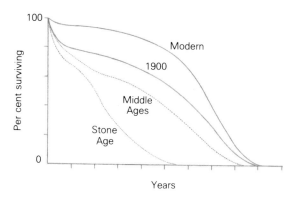

One of the problems of studying ageing is that it is peculiarly difficult to define just what is going on. In part, this is because ageing is not a unitary process and probably involves many factors. For example, we can quite easily understand the ageing of teeth—it is simply a question of wear and tear. In fact, one of the reasons elephants die of old age is because their teeth wear out and are not replaced, and so they are unable to eat. This failure to repair is an important feature of ageing. Wear and tear on joints similarly takes its toll with age. The failure to repair is also very important in the ageing of the brain where there is a continual loss of cells from quite early on in life. Neurons cannot multiply and since no new ones are generated the total number of nerve cells in the brain gradually declines. While cancer and diseases of the heart and vascular system can also be looked at, in part, as a failure to repair, their causes are more complex. Cancer is, as we have seen, due to the accumulation of mutations in the DNA. Some cardio-vascular diseases result from the accumulated damage—in this case a series of events that eventually lead to the blocking of blood vessels and so to failure of the blood supply to a particular organ, particularly the heart. In the case of stroke, there is rupture of blood vessels and subsequent damage to the surrounding brain tissues.

The disposable soma theory of ageing leads us to expect to find evidence of more investment in repair mechanisms in those animals that live a long time and reproduce late, than in those that have a short lifespan. Where should one look for this evidence? One key system is that involved in DNA repair, for if the DNA is damaged, that damage will be amplified for it will lead, most likely, to the production of faulty proteins. There is a complex system for the repair of DNA in the cell but the evidence that it functions better or longer in animals with a long lifespan remains equivocal.

A possible clue to the nature of ageing comes from the observation that normal cells growing in culture are not immortal. Cells in culture such as those from connective tissue will multiply until they form a monolayer and then stop (Chapter 11). If some of these cells are then cultured at low density they will again multiply until they cover the bottom of the dish. But this repeated subculturing followed by multiplication cannot go on indefinitely. After a more or less fixed number of cell cycles, multiplication ceases. The cells do not die but just stop multiplying. For connective-tissue cells taken from young adults the number of doublings is about fifty, but the older the donor the fewer the doublings. It is tempting to think that this could provide an ideal experimental system for studying ageing and much effort has been given to trying to interpret this phenomenon in terms of an accumulation of errors in the cells, leading ultimately to their inability to multiply. The idea fits in well with the development of children suffering from the rare condition known as Werner's syndrome, in which children age prematurely and a young boy may look like an old man. The fibroblasts taken from such children show a very restricted capacity to proliferate.

A quite different interpretation of the restricted capacity to multiply is that it has nothing to do with the ageing process but reflects the cells' response to being placed in culture away from their normal environment. The cells may have a mechanism to restrict unlimited multiplication which could lead to cancer. The idea of accumulation of error being responsible for stopping multiplication is also very hard to reconcile with the ability of cells to become immortal and multiply indefinitely when infected with certain viruses or treated with chemicals that cause cancer. In fact, becoming immortal seems to be a precondition for developing cancer.

Not many scientists work on the problem of ageing even though its importance cannot be doubted. Ageing thus provides a nice example of science being, as Peter Medawar put it, *The art of the soluble* (1967). The problem is, at this stage, just too difficult, and gifted scientists with their skill in choosing the right problem, recognize this. As yet, we do not even understand the timing of puberty or the menopause. Only when an experimental approach to solving the problem becomes available will more scientists enter the field. Perhaps when the nature of the developmental programme is fully understood such an approach will become possible.

REGENERATION

REGENERATION IS closely related to embryonic development. Newts can regenerate complete limbs and this involves cellular activities similar to those involved in the development of the limb in the embryo. But the processes are not necessarily identical. In the case of limb regeneration the new limb comes from adult cells at the cut surface, rather than from a limb bud. What is remarkable is that from the adult tissues a structure resembling the limb bud develops and that regeneration may only replace the missing part. Other animals, such as hydra, also show remarkable powers of regeneration when parts are removed, and this resembles some of the regulative properties of early embryos (Chapter 3). Taken together, many aspects of regeneration seem related to embryonic regulation and can be considered in terms of replacing those positional values of cells that have been lost.

Gradient theories, which are the basis of positional information, had their origin in attempts to understand regeneration. The idea of gradients was first introduced at the end of the nineteenth century by the American embryologist Thomas Hunt Morgan. Studying the regeneration of worms and marine hydroids he was struck by the different rates at which regeneration occurred when an animal was cut at different levels—the further away from the original head the slower

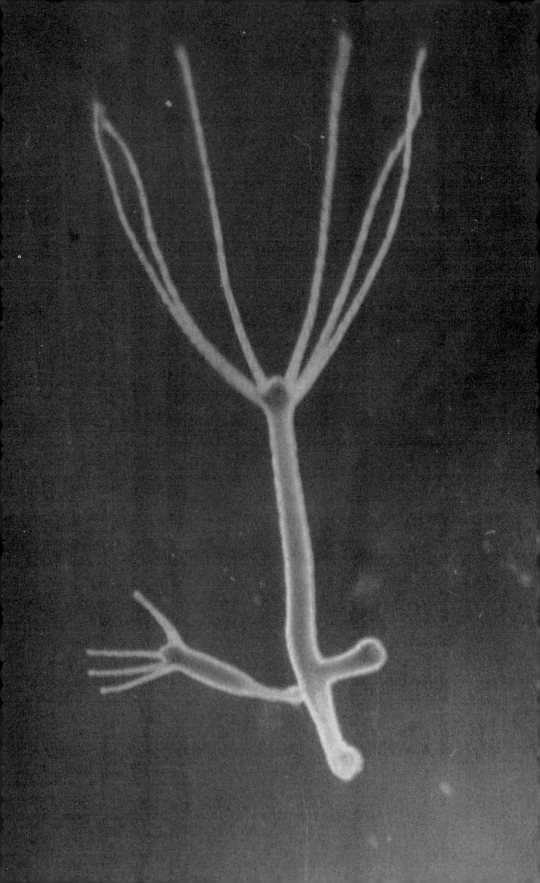

the regeneration. In order to explain his results he suggested that there was some graded property along the animal and that this gradient determined both the polarity of the system and the rate of regeneration. But although his analysis was by far the clearest and closest to our current thought, he had an inexplicable ambivalence towards gradients. For example, he would, at intervals, abandon gradients and return to another model, poorly defined, based on mechanical tensions. As mentioned earlier (Chapter 7) he may have found development and regeneration too difficult and so turned to genetics using *Drosophila* as his model system. Advances in genetics owe much to his change in direction and he was rewarded with a Nobel Prize.

H Y D R A

Not many animals can survive decapitation, but hydra not only survive but regenerate a new head. One can cut up the animal in numerous different ways, and within a few days a hydra with the normal form will be restored. A fragment as small as one-twentieth of the body will regenerate into a tiny well-proportioned hydra with tentacles, mouth, and foot. Fragments of the body column can even be threaded on to a fine hair to make, to begin with, excessively long animals. But, again, within days, heads and feet will form at more or less regular intervals along the tube and individual hydra eventually separate out.

When the ability to regenerate was first announced in 1744 by Abraham Trembley, a Swiss naturalist, it shocked the scientific world since it dealt a blow to preformation theories. Hydra is a small glove-shaped animal with tentacles for catching prey at one end and a sticky foot at the other, and Trembley was fascinated by this small creature which he found in a local pond. He placed a hydra in a drop of water in the palm of his hand, and, with a fine instrument, cut off its head and returned the headless hydra to a dish. To his amazement and delight a new head had formed within a couple of days and the hydra was able to function quite normally. For the preformationists this ability to regenerate lost parts presented a problem, for, if everything was preformed in the egg where could the new parts come from? But conviction can drive the imagination and models were put forward which proposed, for example, that there were latent preformed 'head units' in the animal which became activated when the head was removed.

Hydra thus shows similar regulative capacities to those seen in early embryos (Chapter 3). The pattern in hydra is the same over quite a large range of sizes and the polarity is maintained—in isolated pieces heads regenerate at the surface closest to the original head end. These

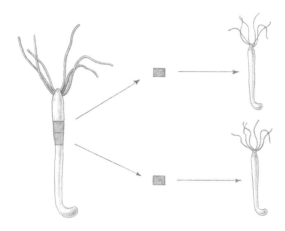

properties were the original stimulus for the French flag problem since, as a convenient approximation, the hydra can be thought of as having three regions, head, body, and foot. The problem was, how a line of cells making a French flag pattern—one-third blue, one-third white, and one-third red—would always look like the flag no matter which regions were removed or how long the line was. Clearly, the cells needed to change from one colour to another, for regeneration in hydra does not involve cell growth or multiplication. Regeneration involves changing the state of the cells and remodelling the tissue. For example, when the head is removed, the tissue just beneath changes into a new head region.

An important aspect of regeneration in hydra is that it does not involve growth. Head regeneration will occur even when all cell multiplication is prevented. It involves a remodelling of the remaining tissue and growth only occurs later when the animal starts feeding again.

The possession by the cells of positional information provides a formal explanation for the regeneration of hydra. If the cells have their position specified with respect to the head and foot ends, then, in principle, the cells 'know' where they are in the animal and can behave appropriately. Direct evidence for the head's boundary-like properties

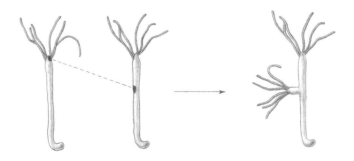

comes from its ability to organize a new hydra axis. When a small fragment of the head is grafted into the body of another hydra a new hydra, with a head, is formed. The head thus behaves like Spemann's amphibian organizer (Chapter 3). In these terms, removing the head removes the boundary or reference region with respect to which the cells have their position specified, and so the key event in head regeneration is to establish a new head end boundary. The way this is done depends on two gradients.

The head in hydra produces an inhibitor which diffuses down the body and prevents any of the other tissues from making a head. When the head is removed the concentration of the inhibitor falls and a new head can now be made. Both the inhibitor concentration and the capacity to make a head are graded and are highest at the head end and decrease with distance from it. So, when the head is removed the greatest fall of inhibitor is at the head end which can then differentiate to form a head. Once the head has regenerated the gradients are re-established. As might be expected there is a similar system operating at the foot end. The regeneration of hydra thus essentially involves two separate processes—the one is to establish boundary regions at the head end and the foot end, and the other is to specify cell position with respect to these boundaries. Viewed in this way it is an elegant example of a solution to the French flag problem.

Has regeneration in hydra evolved as an adaptive mechanism to restore lost parts? Quite the contrary. Regeneration of hydra is not an adaptive mechanism for restoring lost parts—the loss of a head is a rare event in the life of hydra. The regenerative and regulative powers of hydra are due to its normal mode of reproduction which is by budding. About half way down the body column buds start off as small protrusions. These protrusions extend and then tentacles form at the end, while a foot forms at the site of attachment to the main body and a small new hydra detaches itself. Since a whole new hydra is remodelled and patterned from tissue in the adult body column, budding is essentially the same as regeneration.

GROWING A NEW LEG

Growth, by contrast, is central to the regeneration of a newt's limb. When a newt loses a limb the wound heals over and beneath the tip, over the next week, there is a gradual accumulation of cells which form what is known as the blastema. Over the following week the blastema at the tip grows and differentiates and gives rise to that part of the

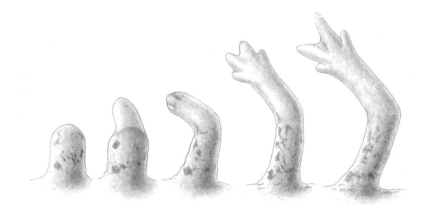

limb that was lost. If the level of amputation is in the middle of the humerus then the missing half of the humerus and the rest of the limb will regenerate; if the level of loss is at the wrist then just a hand will regenerate.

The problem then is how the blastema 'knows' which structures to make and to, apparently, only give rise to the missing parts. Do the cells have access to global information about what structures are present in the limb? The evidence is that they do not have such information and their behaviour is totally determined by very local events at the cut surface. Regeneration can be thought of as generating new positional values distal to the cut surface. The cells along the limb have positional values that were set up during embryonic develop-ment (Chapter 4). The blastema contains cells whose positional values correspond to those at the cut surface and the blastema is somewhat analogous to the progress zone at the embryonic limb. That is, it undergoes growth and generates more distal positional values starting at the level of the cut surface. In this way continuity is established at the cut surface and only missing distal structures are regenerated. For example, a cut at the level of the humerus would give a blastema with proximal positional values, whereas a blastema at, say, the wrist, would have much more distal values. This explanation does not require the cells to have global knowledge of the limb. The cells in the blastema merely generate more distal positional values. This will replace missing parts with normal regeneration, but under some circumstances it can be shown that structures will regenerate even though they are already present.

A classic experiment in the 1930s showed how it was possible to so arrange things that the limb can regenerate from both proximal and distal cut surfaces. The hand of the limb was inserted into a flap in the newt's skin on its under surface and a blood circulation established. The limb was then cut through the humerus and there were now two surfaces from which regeneration could take place. The normal proximal surface regenerated distal structures in the normal way. But what about the distal surface? Would it regenerate the rest of the newt? What it regenerated was exactly the same as that regenerated from the other surface—more distal structures. But the key point is that it regenerated structures that were already present, such as the radius and ulna, even though they were pointing in the wrong direction. Clearly the cells do not make use of information from other parts of the limb but just regenerate in a distal direction.

The leg of the cockroach, which is very good at regenerating, can be used to make the same point about local interactions, even more clearly. The tibia is one segment of the cockroach leg. We can mark different positions along the tibia from say position 1 to position 10 and consider these as representing different positional values. Removal of positions 2 to 9 and joining position 1 to position 10 results in regeneration of the missing positions 2 to 9 and a normal tibia

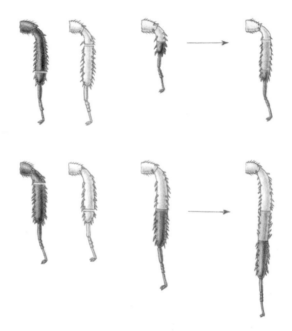

results. If now the tibia is cut at position 10 and the tibia from another cockroach which has been cut at position 1 is joined to it, we again have position 1 abutting position 10, but in this case we have an over-long tibia. To restore normality the tibia should now get shorter eliminating the surplus tissue. Quite the opposite occurs: the tibia gets even longer inserting again positions 2 to 9. Clearly the cells are not responding to any global cues relating to the overall length of the tibia; rather they are responding to local cues.

Here we have the beginning of an important rule controlling regeneration. When cells are placed next to other cells which are not their usual neighbours, they respond by interposing the missing positional values. The system regulates to ensure a continuous smooth sequence of positional values. Thus, when positional value 1 is placed next to level 10 then values 2 to 9 are generated; and it makes no difference in which order the 1 is joined to the 10. These experiments provide excellent evidence for cells having positional values.

A more complex model which deals with the regulation of positional values in two dimensions can account for some surprising results. For example if the left limb of a cockroach is grafted to a right stump then two additional limbs grow out of the junctions, because conditions are established for the generation of more distal positional values. Similar results are obtained with newt limbs. What is remarkable is that the same principles govern the regeneration of the limbs of both amphibians and insects, suggesting that a primitive and fundamental mechanism is involved.

This more complex model considers positional values not only along the main axis but also around the circumference. Around the circumference of the limb the positional values are arranged in a circle like a clock-face running 12, 1, 2, 3 . . . 9, 10, 11 and then again to 12. The values go round continuously. One of the main features of the model is that, unless there is a complete circle of positional values, generation of distal positional values will not take place. So, if a limb is constructed of say, two outer halves, giving 12 . . . 3 . . . 6 . . . 3 . . 12, then regeneration fails. A further feature is that when there is a complete circle of values distal transformation always takes place. It is this rule that accounts for the additional limbs growing out at the junctions when a left limb is grafted to a right stump.

The structures that the newt limb regenerates depends on the positional values at the level at which the limb has been amputated. Retinoic acid has the unique ability to change the positional values in the blastema. If a cut is made at the level of the wrist and it is treated with retinoic acid, the positional values are altered to those at the level of the shoulder. The regenerate formed is no longer just a hand but a complete new limb regenerated from the wrist. This gives a long double limb which has a humerus, radius and ulna, wrist, humerus, radius and ulna, wrist, fingers. Retinoic acid has altered the positional value of the wrist cells to those of shoulder cells. Again, as in the developing chick limb, (Chapter 4) retinoic acid is somehow involved in specifying positional values.

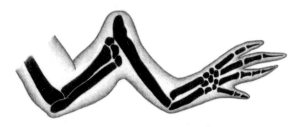

It may be thought that by knowing that retinoic acid alters positional value, we should be able to discover the molecular basis of positional value by an apparently relatively straightforward research programme. Why not just investigate the chemical changes that occur when retinoic acid is applied? In principle, the argument is correct, but in practice it is very much more difficult. Cellular responses always involve many chemical changes and the problem is to know which ones to investigate. It is even very difficult to find out which are the key genes which are activated when retinoic acid is applied. In general, the pathway from signal through early responses to later responses and cellular behaviour is complex, and as yet very rarely worked out in any system.

A peculiarity of newt limb regeneration is its dependence on nerves. If the nerves going from the spinal cord to the limb are cut the limb will not regenerate. The nerves are only necessary for growth, and play no role in determining the nature of the regenerate. It does not even matter very much what sort of nerves enter the limb so long as they are present in sufficient quantity. Their role seems to be to provide the growing blastema with an essential growth factor.

Newts can regenerate other organs, such as the entire lower jaw, and even if the lens is removed the iris will regenerate a complete new lens. Again, as with hydra, one should not think of the newt having these remarkable powers to adapt to environmental traumas; rather its regenerative powers should be viewed as a fortuitous continuation of the ability of embryos to regulate when parts are removed. The adult newt has retained some embryonic characteristics.

REGENERATION AND REGULATION

Mammalian limbs do not regenerate. Nevertheless, repair processes are occurring continually in the normal life of mammals including man. Replacement of cells is a normal event in blood, skin, and the lining of the gut (Chapter 6). Mammalian organs, such as skin and liver, are capable of undergoing repair when damaged, but these processes are not necessarily similar to those occuring in embryos. There is, however, some evidence for the capacity for limb regeneration since young children can regenerate the tips of their fingers so long as the cut is not below the first joint. It would be a major achievement to understand why mammals do not have the ability to regenerate and even more so if that ability could be restored. At this stage we have no satisfactory explanation other than that mammalian limbs no longer seem able to generate cells with new positional values: they have lost that embryonic character. But until we know what this means in molecular terms it is not a very useful explanation.

The mechanisms involved in regeneration are very similar to those involved in regulation of the early embryo when parts are removed or rearranged (Chapter 3). Many aspects of both regeneration and regulation can be understood in terms of the generation of new positional values. This suggests that the processes represent a fundamental aspect of embryonic development, a view that is strengthened by the similarity in the processes in animals as different as insects and amphibia. Further support for the basic nature of the processes comes from a surprising source.

The protozoa are single-celled organisms which can be very complex. One large group is the ciliates. As their name implies, they have cilia on their surface which are small whip-like structures which enable the animal to swim and direct water into the mouth. The pattern of cilia and other surface structures can be very complex in these single-celled animals. The remarkable feature is that they are capable of regeneration and regulation when parts are removed, and that the rules governing this regeneration are very similar to those

found in insects and amphibians. It is, for example, as if there is a set of positional values covering the surface of the cell and there is a tendency to fill in missing values. It is extraordinary that the same phenomena can be observed in both a single cell and multicellular system. One cannot but believe that there is some basic mechanism at work which is not yet understood.

EVOLUTION

T HE STORY is told of the dilemma that faced the great embryologist Karl von Baer. 'I have' he wrote in 1828 'two small embryos
preserved in alcohol, that I forgot to label. At present I am unable to
determine the genus to which they belong. They may be lizards, small
birds, or even mammals'. Karl von Baer spent much time looking at
the embryos (labelled!) of different animals and drew some important
general conclusions. He noted, for example, that the common feature
of animals appeared quite early in development and that specialized
features only appeared later. For example, the heads and bodies of
early vertebrate embryos are, as von Baer was forced to recognize in
his unlabelled specimens, very similar, and only later do the distinctive features of fish, bird, and human heads and bodies emerge. Here
we have a very important relationship between evolution and
development, of which von Baer was unaware, yet which was clearly
stated by him, namely that the general features of a group of animals,
like the vertebrates, appear early in development and from these the
more specialized characters develop.

This relationship between general and specialized characters can be
understood in terms of evolution. During evolution, embryonic
development is modified so as to give different characters to the adult

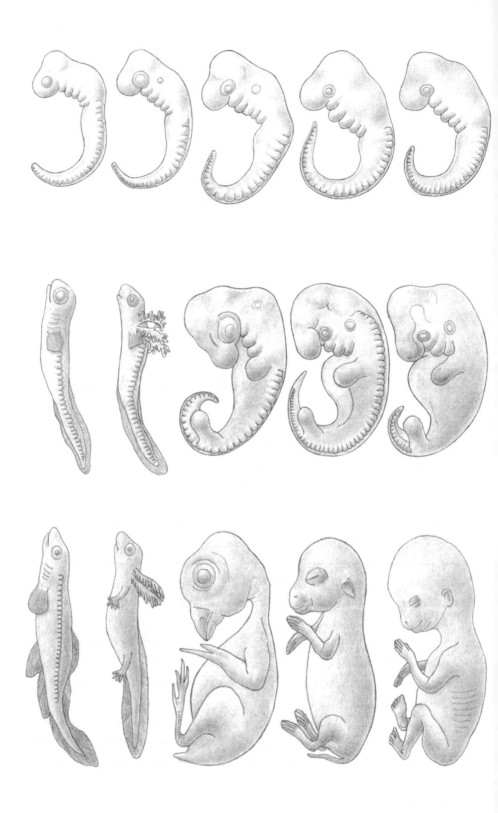

animal. Those animals that have evolved later are the result of the modification of the development of their primitive ancestors. Quite often, some of the characters of the embryo of the primitive ancestor can still be seen. An example is the arches and clefts in the head region of the embryos of all vertebrates. These arches and clefts gave rise to the gills in primitive jawless fish. In later vertebrates they became modified to form jaws and other structures. A higher animal, like a mammal, passes through an embryonic stage when there are structures that resemble the gill clefts of fish. But this resemblance is illusory and the structures in the mammalian embryo only resemble the structures in the *embryonic* fish that will give rise to gills.

Ernst Haeckel, who was Darwin's chief supporter in Germany, took a very different view. Haeckel claimed that the embryo passed through the *adult* stages of its ancestors and thus a study of embryonic development could reveal how animals evolved. He coined the phrase 'ontogeny recapitulates phylogeny' to summarize his famous, or rather infamous, law. Ontogeny merely means embryonic development and phylogeny is the evolutionary history of an animal. Further, he regarded man, *Homo sapiens*, as the pinnacle of evolution and the successive stages of development that the human embryo passed through therefore corresponded to the adult forms of lower organisms from which humans had evolved. For Haeckel, that stage in human development in which there are a set of arches and slits behind the head represented a primitive fish ancestor, and not, as von Baer made quite clear, structures present in embryonic fish. In retrospect, it is not easy to understand why this theory should have received such wide support. Even Freud was greatly influenced and his ideas on the id and ego and stages in psychic development reflect Haeckel's view. Embryos, however, do not pass through the adult stages of their ancestors; ontogeny does not recapitulate phylogeny. Rather, ontogeny repeats some ontogeny—some embryonic features of ancestors are present in embryonic development.

In any group of related animals, such as vertebrates, there is a stage in development which is common to all members of the group. This stage is called the phylotypic stage and all members of the group will have a developmental programme that takes them through the phylotypic stage. For vertebrates, the phylotypic stage immediately follows gastrulation when the main body axis with a very primitive head, and the first few segmented somites, can be seen. Before and after the phylotypic stage the developmental pathway may be quite different. For example, the early development of vertebrate embryos and the way they gastrulate may have only some features in common, although at the phylotypic stage—the one that von Baer found difficult to recognize in his unlabelled bottles—they look very similar. A mammalian embryo follows its own novel developmental pathway up to the time of gastrulation: the early events generate tissues that will form the extra-embryonic structures which are involved in implantation and placenta formation, whilst the embryo proper develops from a much smaller group of cells (Chapter 3). By contrast the amphibian embryo develops directly from a very yolky egg. In a way, the phylotypic stage is the key feature which links all vertebrates. For reasons we do not really understand, it seems to be quite easy during evolution to modify development before and after the phylotypic stage, but that stage itself remains constant. It is an essential stage for the development of a functional animal; perhaps just because it permits earlier and later diversity.

A nice example of modification of an organ during evolution comes from those very arches—the slits that Haeckel thought represented the gills of an ancestral fish. In the distant past, they did indeed develop into the arches which carry the gills in adult fish and provide the means for oxygen exchange. The primitive fish had no jaws and the anterior arches became modified to form jaws in what was undoubtedly a major evolutionary advance. Our own jaws still develop from the first arch that appears during development of the head region. In fish, the more posterior arches and clefts still develop into gills; in all mammals, however, these arches are only transitory structures but give rise to structures in the throat. When, during evolution, the ancestral fish left the sea and gave rise to animals that could live both in water and land— the amphibia—these arches and clefts were no longer needed. Nevertheless, they continued to persist in embryonic development even though they no longer formed gills. They provided, instead, a system which could be modified to provide some other adaptation for the animal. The evolutionary process could now, to use the French molecular biologist, François Jacob's phrase, tinker. Tinkering involves using materials for purposes quite other than that for which they were made. Tinkers make use of whatever is at hand.

Another example of tinkering is the origin of one of the three small bones that transmit sounds from our eardrum to the inner ear. In our lizard-like ancestors—the reptiles—there were only two bones in the chain, which provided an effective but less efficient transmission. Where did the third element come from? It came from the lower jaw. Unlike mammals, which have but a single bone in the lower jaw, our reptilian ancestors had several bones in the jaw. When, in evolution, there was a change to a single bone, which had the advantage of greater strength, one of the jaw bones became redundant. This bone, which was now at the back of the jaws, through tinkering, took on a completely new function, becoming the third bone in the sound transmission system.

Lest any one should doubt their fishy past the evidence from the development of the kidney should be compelling. All mammals start developing a kidney very similar to that which develops in a fish. It develops first at the anterior end of the body cavity but its development is transitory and disappears. A little further back along the body another fish-like kidney develops, but it too is not functional. It does not disappear but becomes modified to form structures associated with the ovary and testis. The functional kidney develops in a yet more posterior position from quite different tissues.

VARIATIONS ON A THEME

Changes in the developmental programme can only be brought about by changes in genes. Whatever tinkering achieves, it can only do so by changing the genetic constitution of the embryo. The changes take place over a very long time and represent the result of many small changes. This gradualist view stands in contrast to those who have suggested that quite dramatic changes could be brought about in just one generation by the appropriate combination of genes. There is no evidence for such a mechanism. An examination of, for example, the evolution of the horse's leg provides good evidence for gradualism and how fewer changes in genes than expected might be required. Simply getting bigger can have profound results.

The basic pattern of the vertebrate limb has been modified to give limbs as varied as those of the bat and the horse (Chapter 4). The origins of the limb itself goes back to the fins of those fish that evolved into land-dwelling animals. While the details of the transitions are fuzzy the similarity between the fins of fossil fish and the basic limb pattern is clear. It is this original pattern that has been retained and modified.

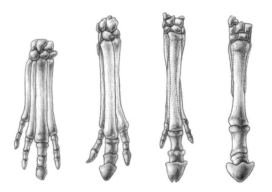

Modern-day horses run on single toes whereas their ancestors had five digits firmly planted on the ground. The horse's leg has become modified so that the central digit is long and strong and ends in a hoof, while the digits on each side have virtually disappeared. All that is left are two small splint-like structures which never make contact with the ground and would be totally ineffectual even if they did. Fossils provide the evidence for this transition from five-toed horses to essentially one-toed horses.

As the horse became larger during evolution there was a tendency for its legs to get longer and also for the animal to run on its toes. For, because of the relative growth rates, the central digit, whose growth rate is faster than the lateral digits, will become proportionally much longer. Initially, just three toes made contact with the ground and then, as the horse further increased in size, the central toe became proportionately so long that even the two side digits no longer made contact with the ground. It does not require great imagination to see how small changes in growth control could lead to the reduction in size of the lateral digits, and to a further increase in length of the central digit. This change in limb structure may thus require less genetic change than might be supposed. In part, it may have resulted from horses simply getting bigger. Further local changes in the growth rates could have reduced the lateral digits to splints. The genetic changes required to alter the developmental programme may be relatively few in number.

The Irish elk may have died from excessive growth. Fossils of this enormous deer show that it stood nearly 10 feet tall and had gigantic fan-shaped antlers with a span of up to 12 feet. These elks became extinct about 11 000 years ago and the speculation is that they were literally weighed down by their gigantic antlers. The further speculation is that the over-large antlers developed in the elk because of the relationship between antler growth and body size. Analysis of fossil skulls suggests that the antlers grew at a rate of two-and-a-half times faster than the rate at which the skull grew. So, as the animal increased in size, which we presume had some adaptive value, the antlers, with their intrinsically rapid growth rate, increased their size even more. They became too large to be supported.

While increase in size has important effects on body form, it is the local growth rates that are crucial. Changing local growth rates can bring about dramatic changes in the shape of shells. Shells often grow as regular spirals when viewed along the main axis. In addition, they grow along this axis giving them a three-dimensional spiral form.

Computer simulations of shell growth have been carried out, and by varying the numbers that control growth rates in the different directions it is possible to generate many of the shell types that are representative of different mollusc groups. The important point in these simulations is that the changes in form do not involve changing the basic growth mechanism but merely the different rates at which growth occurs in various regions. These could be under the control of genes. The situation is very similar to the way in which small changes can alter limb growth or even the form of folding cells sheets (Chapter 2). Thus changes in genes controlling growth rates can, in principle, convert one kind of mollusc into another.

The British biologist D'Arcy Thompson's great book, *Growth and form*, was written at the beginning of the century, when very little was known about mechanisms of development. However, he did have brilliant insight into coordinate growth. Thompson pointed out that if the overall shape of an organism such as the body of a crab or a fish is plotted on a rectangular coordinate system, then by smoothly distorting the coordinate system, the body shape of related fish or crabs could be quite easily derived. It was like drawing the shape on a flat sheet of easily deformable rubber and then stretching or compressing

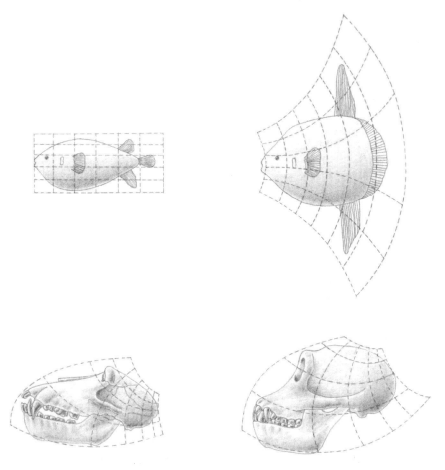

the sheet in different regions. All the transformations were smooth and continuous. The transformation is, of course, not in the coordinate system, but in the local growth rates at different positions within the same coordinate system. Similar transformations could be drawn for skull shape. The face (Chapter 2) develops from a series of bulges in the head region and at early embryonic stages it is not easy to

192

distinguish dog from cat, mouse from man. The differences in facial features are very dependent on just how much these bulges grow. One can begin to imagine how genes could control such changes in growth rates at different positional values.

The key changes in the evolution of form are in those genes that control the developmental programme for the spatial disposition of cells. The difference between chimpanzees and humans lies much less in the changes in the particular cell types—muscle, cartilage, skin, and so on—than in their spatial organization. Direct confirmation for this comes from studies which compare the proteins of humans and apes. If we look at the genes that code for the average 'housekeeping' proteins—proteins that function as enzymes or provide the basis for cell structure and movement—the similarity between chimpanzees and humans is greater than 99 per cent. The differences must reside not in the building blocks but in how they are arranged, and these are controlled by the regulatory genes controlling pattern and growth.

A NEOTENOUS APE

Alterations in the timing of events in development have brought about important changes. For example, the newborn kangaroo has very well developed arms with which it can climb to the pouch, while the legs are still rudimentary. Among the most interesting changes in timing are those in which there has been a retention of 'juvenile' characters while the animal becomes sexually mature—a process known as neoteny. Effectively, the time of sexual maturity has been brought forward. For example, salamanders, like the Mexican axolotl, do not progress beyond their larval form, but become sexually mature without undergoing metamorphosis. The larval salamander is already quite advanced, having well-developed limbs, but its most striking larval features are large gill-like structures through which it breathes in water and which confines it to living in water: like fish, they cannot breathe on land.

In *After many a summer* (1939) by Aldous Huxley, a scientist has been hired to discover an elixir which will give immortality to whoever drinks it. He eventually hears about a gentleman of noble birth who has, for many years, been experimenting along similar lines. The man is now over two hundred years old. When the scientist visits him he finds him in good health, but that he has become a gorilla. The scientist concluded that with continued growth the 'foetal anthropoid was able to come to maturity'. Aldous Huxley was the half-brother of the zoologist Julian Huxley, who worked on relative growth rate, and knew that his story was more than just a joke. A plausible story, for the origin of humans was that they were premature apes, that they had descended from immature ape-like animals which became sexually mature.

Human beings may well have evolved from neotenous ape-like forms. There are a number of features which would support such a view, such as a reduction of body hair, a high relative brain weight, the persistence of growth of the skull for some time after birth, and small teeth. All these characters are found in juvenile apes. If this is true, being human may be the result, in part, from changes in those genes that control the timing of developmental processes. In general terms, the retention of juvenile characters may have been important for evolution. As Stephen Gould puts it, neoteny would 'provide an escape from specialization. Animals can slough off their highly specialized adult forms, return to the lability of youth, and prepare themselves for new evolutionary directions.'

DEVELOPMENTAL CONSTRAINTS

A way of looking at changes in evolution lies in the developmental mechanisms themselves. For example, widening the developing limb bud can result in the appearance of an additional digit (Chapter 4). There is no requirement for complex genetic changes to specify this additional digit. Note, too, that the digit is either present or not and there are no intermediate forms. In a way, making an extra digit, or losing one, is very 'easy' for the embryo and it is just these 'easy' developmental changes that could provide the basis for change. Again, the folding of sheets of cells, so fundamental to the moulding of form, seems, for the embryo, easy, and may not require complex genetic changes. And another example of apparent facility is the manifestation of Crick's aphorism that embryos are very fond of stripes. Breaking up a region into evenly spaced patches comes, perhaps, easily to the embryo.

If certain changes are easy for embryos, others are very difficult. It is easy to make an extra digit but what about an extra humerus or a whole extra limb? Or, to consider a bizarre possibility, could liver cells or an eye develop at the tips of the fingers? Could these developmental constraints be an important controlling factor in evolution?

Developmental constraints are raised by the problem posed by the American evolutionary geneticist Theodor Dobzhansky. Could, he asked, human beings ever evolve into angels? He concluded that it would be possible, with appropriate selection, to evolve an angelic disposition but that it would not be possible to evolve a pair of wings in addition to the arms. To put it another way: given as much time as you like—up to say a thousand million years, which is about how long multicellular life is thought to have existed—and as many people as you like, would it be possible to design a breeding programme which would result in, if not angels, winged humans? Could mutation and selection lead to wings, which requires both the development of an extra pair of arms, and feathers? The more general question is whether embryos can make all imaginable forms. Are some forms much more difficult to make than others? Are there constraints which development puts on evolution so that certain forms are effectively unattainable?

There is, in principle, an easy solution to the angel problem but it really is cheating. If one knew enough about what genetic changes we required then all one need do is to examine the DNA of all people and select those in which changes in the DNA were proceeding in the desired direction. This is cheating because it is privileged knowledge which is only obtained by examining the DNA and selecting at the level of the genes rather than that of the organism. Selection in evolution acts, not on the DNA, but on the animal. Most of the changes in the DNA will have deleterious effects or not cause the necessary changes and so there is no way of selecting the desired changes. Consider, for example, an extra arm. There is no way of selecting for it since there are no intermediate forms; it is not like selecting for increased arm length where one can select for small increases. It is all or none. The chance of an extra arm developing because of mutations is very very small—effectively negligible. The reason it is so small is that it requires the coordinate change of many genes and the chance of these occurring simultaneously is virtually zero. Similarly, to make feathers instead of hairs requires a significant number of changes in the genes and there is no way of selecting for the intermediate stages. By contrast, making hairs longer or shorter or thicker is very easy because these properties vary continuously. The same holds true for angelic disposition.

So, developmental mechanisms, together with their genetic control, put a severe constraint on the evolution of animal form. It is not selective pressures that have kept the basic pattern of the vertebrate arm the same, but the fact that altering the basic pattern is almost impossible. Therefore, not all imaginable animals are possible. Any theory of evolution must incorporate an appreciation of developmental mechanisms.

EVOLUTION OF DEVELOPMENT

Given the 'miraculous' evolution of the cell, there is the problem of how the embryo, and development itself evolved. What was the origin of nature's triumph, the embryo? What new features needed to evolve in order for embryos to make multicellular animals, and what was the origin of, for example, embryonic processes like gastrulation? Given the cell, it seems that there was no new major step required, no brilliant new invention. All it required was to make use of the properties that cells already had and marshall them in a slightly different way.

Development requires a programme of gene activity, spatial organization, and cell movement. All these were already present in the primitive cell. The cell's capacity for multiplication already contained most of the features required. In the cell cycle involved in cell multiplication (Chapter 11) there is a well-defined genetic programme that takes the cell through the successive stages of the cycle, and these stages can be thought of as different states of cell differentiation, because specific sets of genes are switched on and off and specific proteins are being made. Also, at cell division there is a well-patterned spatial organization which distributes the chromosomes and divides the cell in two. Movement, too, was already present. It was, of course, necessary for cells to remain together and communicate, and while not underestimating the problems involved in the transition from single cells to multicellular organisms, there is no reason to believe that a great deal new had to be invented. Moreover, some single-celled animals, like the ciliates, had already probably evolved a remarkable capacity for generating spatial organization (Chapter 13).

Given the transition from single cell to the multicellular state, there still remains the problem as to how embryos themselves evolved. Consider gastrulation. How could the quite complex movements that take place during gastrulation have evolved; and what was the selective advantage for such a process? For the answer we turn back to Haeckel. Although he was quite wrong about ontogeny recapitulating phylogeny, he did provide within that framework a very plausible mechanism for the evolution of the gastrula. He thought that the gastrula stage represented the adult stage of a very primitive ancestor, the Gastrea. With some modification this makes a great deal of sense.

The earliest multicellular organisms were little more than a hollow ball of cells. All that was required for their development was cleavage to divide up the primitive egg. The next evolutionary stage may have been the entry of some of the cells from the wall into the interior where they might have provided some specialized function like digestion. But the crucial step was the formation of a special feeding region which was the forerunner of the primitive mouth and gut. Imagine, with Haeckel, the ball of cells living on the ocean bottom and feeding from the surface on which it rested. Some of the cells in contact with the surface could have become specialized for feeding, and catching smaller single-cell animals. Perhaps then there would have been a small inward folding of the sheet to help catch the prey. Here was the beginnings of a primitive gut, an infolding specializing in feeding. This was Haeckel's primitive Gastrea and the possible origin of gastrulation. Little further was required to make the infolding more extensive so that there was now an organism with a primitive gut but as yet no mouth. This Gastrea would have been very like the larval stage of many hydra-like animals and the early stage of sea-urchin gastrulation (Chapter 2): those primitive animals would have been like a stage in gastrulation, as it now occurs. It is not known if this story is correct but it does show how gastrulation could have evolved.

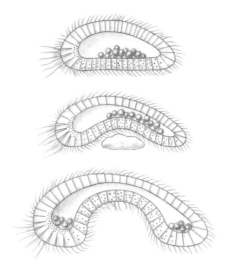

Development has been rather unadventurous over the last thousand million years. Unadventurous in the sense that it is unlikely that any new developmental mechanisms have evolved. Yet the diversity of animal life is astonishing. The basic mechanisms of differentiation, patterning, and moulding of form are probably unchanged. The triumph of the embryo has been to exploit these basic and primitive mechanisms with brilliant success.

A PROGRAMME FOR DEVELOPMENT

N O ONE who studies development can fail to be filled with a sense of wonder and delight. Development of the embryo is a truly remarkable process. Understanding the process of development in no way removes that sense of wonder. Of course, we only partly understand how embryos develop, but we can see, in outline at least, most of the main themes. General principles seem to be emerging, but there is still an enormous amount of work to be done. Most of the difficult problems require an understanding of how cells work and how genes control these activities.

The embryo does not contain a description of the animal to which it will give rise, rather it contains a generative programme for making it. A generative programme is like a recipe, it is quite different from a descriptive programme and a quite complex form can come from a simple programme. The generative programme is essentially contained within the genes, and since genes exert their influence by controlling what proteins are made, development can, in a sense, be thought of in terms of the specific proteins that control cell behaviour. For, ultimately, it is cell behaviour that provides the link between genes and pattern and form. There are thus no genes for 'arm' or 'leg' as such, but specific genes which become active during their formation. The complexity of development is due to the cascade of effects, both within and between cells, when the synthesis of particular proteins is changed.

There is no 'master-builder' in the embryo. Each cell in the developing embryo has access to the same genetic information. Each cell has the same set of genes which are used for building the animal. To generate differences between cells, the cells 'converse' with each other. Just how complex these interactions are remains to be determined. It could be that the same signals are used again and again and that the different responses reflect the cell's internal programme which, in turn, is determined by its own developmental history. A cell in the developing heart is on a quite different developmental pathway from a cell in the limb, and will respond to the same signal in a different way. In general terms, my guess is that the key signals between cells will turn out to be rather few in number. But that may turn out to be wildly optimistic.

A general principle of embryonic organization is that 'small is beautiful'. There is no central government but rather, a number of small self-governing regions. All interactions between cells, when the basic patterns are being established, take place in relatively small groups of cells, and signals very rarely go beyond about 20 cell diameters. When the system becomes larger it gets broken down into a number of autonomous regions, whose development is largely independent. Embryonic organization cannot, apparently, deal with communication between large numbers of cells at the same time.

It may be possible to classify the genes involved in development into three main classes: those involved in patten formation, those involved in cell differentiation, and those involved in changing shape and in growth. One might also think of development occurring in just that sequence-patterning, differentiation and then change in form. If that were the case, it becomes possible to see how the processes are linked by the successive activation of genes. The patterning genes could be turned on at specific concentrations of a chemical gradient.

Then the product of these genes could bind to the DNA adjacent to the differentiation genes and so activate them. In addition, their products could also activate the genes involved in change in form such as those coding for cell adhesion molecules. In principle, at least, it is a plausible scenario.

There is still a great deal to be discovered about the nature of the developmental programme. It is very hard to know just how far we have to go. Do we have the right general principles, or are there going to be some big surprises that will completely upset the way we think about development? I doubt it. Even so there are many genes—thousands—still to be identified and many years of study will be required before we understand fully how each one controls cell behaviour. In a way, cells themselves provide a major stumbling block, for the internal network of chemical reactions within the cell is very much more complicated than the interactions between cells. Cells are more complicated than embryos. Nothing illustrates this better than the remarkable capacity of single-celled organisms to regulate the pattern on their surface in a manner very similar to that shown by regenerating multicellular animals. Understanding this process could reveal a fundamental developmental principle.

Developmental biology is at a very exciting stage. The advent of molecular biology and the introduction of molecular techniques has transformed classical embryology from what can now, with hindsight, be seen to have been a rather stagnant phase. Not that the classical approaches were without value. They helped to define the problems which we are now able to analyse in molecular terms. Nowhere is the excitement more evident than in the results that have come from studies on the development of the fruit-fly. A combination of genetics, classical embryology, and molecular biology has given us not only the clearest picture of what the developmental programme looks like, but, even more important, has provided tools for identifying key developmental genes in other organisms—from frogs to mice. Those of us who work on vertebrates are in great debt to the fly people.

I have always believed that general principles of development would be discovered. It is too soon to be sure just how true this is. At present one can see gene networks, gradients, and cell adhesion, and movement providing the basis for such principles. Even if the signals between cells are not the same in all embryos they are at least more likely to be different dialects than different languages. But it was beyond my expectations that, at the level of the genes, function would be similar in groups so apparently different as flies and mice. Yet this is the picture that is emerging. The genes that are involved in patterning the early fly embryo have been used to identify similar genes in mice. So for example, genes controlling the posterior patterning of the fly seem to be closely related to those controlling the back end of the mouse. Development has, through millions of years, conserved basic mechanisms.

It is not necessary to justify the study of embryonic development, for development is absolutely fundamental to all of biology. However, it is reasonable to ask what benefits might come from such studies. In spite of great progress in understanding, so far the practical benefits have been very limited. The best examples are in relation to *in vitro* fertilization and the prenatal diagnosis of genetic diseases. An understanding of development was essential for progress in these fields, even though the knowledge required was restricted to the earliest stages of development. In the future, there can be no doubt that understanding development will, in turn, give a much greater understanding of birth abnormalities such as cleft lip and spina bifida. Almost one child in twenty is born with some abnormality. We still have no real understanding of why thalidomide causes limb deformities.

Understanding does not necessarily mean that the knowledge will lead to a practical benefit. But it is, however, almost certainly a prerequisite. We must remember, as Paul Valéry said, 'We enter the future backwards'. It would be as unwise to promise great benefits as it would be to deny the possibility. Certainly with the new techniques of altering the genetic constituton of animals—ethical issues aside— understanding how genes control development will be of enormous practical value. And while at the moment we can do no more than explain the inability of mammals to regenerate their limbs by saying that, unlike newts, mammalian limbs have lost their embryonic character and so cannot change their positional value; who knows what possibilities to alter that situation lie in the future? Perhaps sooner than we dare hope, it might be possible, by knowing in detail both the genetic information in the egg and how cells work, to compute just how the embryo will develop. That would be another triumph.

INDEX

abnormalities 20, 203
achondroplasia 152
acquired characteristic, and inheritance 135
activin, and induction 45
adhesion 16
  and developmental programme 24
  and neuronal connections 127
After many a summer (Aldous Huxley) 192
ageing
  and cell multiplication 168
  disposable soma theory 166
  and DNA repair 169
  and evolution 167
alligators, and sex 141
amphibian embryo 18, 40–5, 85
amino acids 80
angel problem 194
animal cap, amphibian 45
animal experimentation 10
animal pole, sea-urchin 35
animal–vegetal axis, sea-urchin 35, 39
antenna, transformed to leg 111
antero-posterior axis 40, 116
antibody technique 100
Aristotle 2, 10, 57
apical ridge 60
arm 59
  similarity to leg 67–8
Art of the soluble, The (Peter Medawar) 169
astrocytes 100
axes
  and eggs 40
  and mouse egg 40

axis, main 116
axolotl 147, 192
axon 121
  pioneer 121

Bateson, W. 111
behaviour
  and development 132
  and sex 142
Belousov–Zhabotinsky reaction 55
bicoid
  gradient in 107
  protein 109
  mutations in 109
blastema 176
blastula 11
blood
  lineage of 93
  and stem cell 94
blood vessels
  and cancer 163
  damage to 67
bones, growth of 149
Bonnet, C. 4
boundary region 38
  sea-urchin 39
brain 18, 119
  hormonal influence on 143
  patterning of 127
  and sex 142
breasts 139
Brenner, S. 51

calcium, and cell adhesion 24
CAM (cell adhesion molecule) 24, 126
cancer
    and blood supply 162–3
    and cell migration 163
    and children 160
    developmental programme of 159
    and failure to differentiate 161
    and genes 163
    and mutations 160, 163
    origin from single cell 160
    reversal of 161
cancer cells 84
    requirement for growth factors 159
    variation in 161
carcinogen 159
carcinoma 161
cartilage 59, 149
cat, tortoiseshell 107
cell
    complexity of response 47
    function of 5
    internal programme of 92
    options open to 47
    social nature of 45
    and sorting out of 25
    types 91
cell adhesion
    and calcium 24
    molecules 24
and neurons 127
    and neurulation 24
cell cleavage 11, 52–3
cell communication 39, 44–5, 53, 63, 67, 200
cell contractions 17
cell cycle 157
cell death 72
    in limb 72
    and nervous system 130
    in skin and gut 97
cell diversification 91
cell division 158
cell divisions, programmed 101
cell fate 40–1
cell fusion 84, 85
cell lineage 50–2, 93
cell membrane 7
cell movement 16, 25, 67
cell migration 15, 25–6
cell multiplication 7, 148, 157
cell response, and history 48
cell signalling 38, 44–5, 53, 63, 67, 200
cerebellum, loss of 129
cerebral cortex 127–8
chemical waves 69
chemistry 10
chemo-affinity, and neurons 126
chemotaxis 27

chimaeras 101
Chomsky, N. 132
chromosomes 77, 101, 112, 116
ciliates 196, 201
    and regeneration 182
cleavage 11
    radial 50
    spiral 50
    and preformation 32
cloning 87
cockroach leg, regeneration of 178–9
communication, cell 39
community effect 45
Cohen, S. 131
co-ordinate system 38
complexity, and internal programme 47
computer models 20
crab, growth of claw 153
Crick, F. 21, 194
critical period, and neurons 132
cytoplasm 7
    and control of gene activity 85
    and fly egg 109
cytoplasmic factors 53
    and muscle 54

Darwin, C. 136, 185
Darwin, E. 137
digits 60
dendrites 121
Descartes, R. 4
determination 42
development
    and evolution 183
    evolution of 195–7
    general principles 201
developmental constraints and evolution
    293–5
developmental mechanics 33
developmental programme 17
    and ageing 165
    and evolution 192
differentiation 7–8, 91
differentiation, and branching pattern 95
diffusion, and cell signals 39
dinosaurs, and sex determination 141
DNA 8, 77–90
    and ageing 168
    codes for proteins 80
    and evolution 194
Dobzhansky, T. 194
dorsal gene 110
dorso-ventral axis 40
    fruit-fly 110
Drosophila (see fruit-fly)
Driesch, H. 33–4, 36

ear, and evolution 187
egg
  and axes 40
  chicken 11
  human 11
  mouse 11
embryos
  all-male 142
  all-female 142
epigenesis 2, 10, 57
epigenetic landscape 91
erythropoietin 96
entelechy 34
enzymes 8, 79
evolution
  and angel problem 194
  and development 183–9
  of development 195–7
  and developmental constraints 193–9
  and developmental programme 186–92
  of ear bones 187
  of gastrulation 195–7
  of horse's leg 188–9
  of limbs 188–9
  and regulatory genes 192
extracellular material 27
eye 19
  connection to brain 124
  fruit-fly 116
  inversion of 124
  lazy 132
  and regulation 41

face 28
fate, determination of 42
fate map 41
feathers 56
female, development of 137–8
fertilization 136
  in vitro 137, 203
filopodia
  neurons 121
  sea-urchin 15–16
finger, sixth 69–70
Folkman, J. 162
forces 19
form, moulding of 11, 28
French flag problem 37
  and hydra 174, 176
Freud, S. 185
fruit-fly 105
  antero-posterior axis 107
  best model for development 112
  cleavage 106
  development of 140
  sex in 140

dorso-ventral axis 110
  and excitement 201
  segments 106

gap-junctions 39
Gastrea 196
gastrulation 12–16
  evolution of 195–7
  and fate map 41
gene 1, 8, 79–80
  activity 92
  and cancer 163
  control of 83, 88
  and development 200
  and gradients 105
  hierarchy of, in fly 113
  homeotic 111–12
  imprinting in males and females 142
  master 93
  networks 84, 90
  order along chromosomes 112
  promotor of 83, 88
  regulatory 102, 105
  switches 88
  transcription 83
general principles 112–13
generative programme 17
genetics 1
genetic diseases 82
genetic engineering 88
  fear of 87
genetic information, same in all cells 87–8
genitals 139
germ cell 135–6
  migration 27
germ plasm, continuity of 135
gills, embryonic 185
glia
  in cortex 129
  differentiation of 99
  and migrating neurons 128
gradient 63–4
  and bicoid gene 107, 109
  chemical 38
  control of pattern 107
  and dorsal gene 110
  and fruit-fly embryo 107
  and genes 105
  in hydra 175
grasshopper limb, and neuron guidance 121
growth 28, 145
  abnormalities 152
  autonomy of 146
  and evolution 189–90
  human 149, 151
  mechanical coordination of 147
  and positional value 154

growth (*cont.*)
  programme of 39, 146, 154
  relative 153
  and shell evolution 190
  standard curves for 151
growth cone 121
growth factors
  and cell multiplication 158–9
  and glia 100
  and growth 150
  and induction 45
Growth and form (D'Arcy Thompson) 190
growth hormone 89, 152
growth plates 149, 152
growth spurt 151
Gurdon, J. 85
gut 12, 41
  lining of 96
  stem cells 96
Gould, S. 193

Haeckel, E. 185, 196
haemoglobin 85
Hamburger, V. 130
harmonious equipotentiality 34
Harrison, R. 147
Harvey, W. 4, 77
head 28
  growth 154
  and neural crest 27
heart 18, 20, 59
hermaphrodites 140
Hippocrates 2
homeobox 115
  and limb 74
  as marker of position 116
  in mouse and fly 117
  and nervous system 129
homeotic genes 111
hoemotic genes, leg and antenna 113
homunculus 32, 34, 53–4
hormone, growth 152
hormones, and sexual characters 137–8
horse, evolution of leg 188
human evolution, and neoteny 193
Huxley, A. 192
Huxley, J. 193
hydra 5
  regeneration of 173–6
  and self assembly 25

immortality
  and cancer cells 160
  and germ cells 135
induction 43–5
inner cell mass 49

instruction, and selection 47
insulin 158
interpretation, and positional information 38–9
Irish elk 189

Jacob, F. 187
jaws, evolution of 187

kangaroo, and evolution 192
kidney 20
  and evolution 187

laser 53
Leder, P. 82
Le Douarin, N. 98
leg 59
  similarity to arm 67–8
lens 19
  induction of 46
leukaemias 159–61
Levi-Montalcini, R. 130
limb 59
  abnormal development of 74
  basic pattern of 68, 73
  and evolution 188
  and genes 82
  growth of 146
  kangaroo and evolution 192
  and mutation 82
  and neuronal pattern 123
  and patterning genes 68
  and positional information 61
  regeneration of 176, 180
  and self-organization 69
  vertebrate 68
lock and key, and neuronal connections 126
lungs 18, 20

male, development of 137–8
Malebranche, N. 4
Malpighi, M. 4
Mangold, H. 43
master gene 93
maturation of cells 92
master builder 5
maternal genes, requirement for 141
Medawar, P. 169
melanoma 162
metastasis 157, 162
microfilaments 19–20
molecules, and cell language 10
molecular biology 9, 115
Morgan, T. H. 105, 171

morphogen 55, 63–4
mouse embryo
    and axes 40
    chimaera 102
    cleavage 48
    patterning 48–9
    polarity 36
    regulation 36, 49
Mozart, W. A. 133
Murray, J. 56
muscle 12
    activation of 85
    genes for 93
    growth of 147
    induction of 45
    limb 70–1
    maturation of 92
    tunicate 54
mutations 2, 81–2
    and cancer 163
    and evolution 194
    homeotic genes 111
    limb 82
    nematode 52
myogenin 93

MacBride, W. 67

nematode
    cell interactions 53
    cleavage 52
    lineage 51
neoteny 192
    and human evolution 193
nerve growth factor 131
nerves, and limb regeneration 180
nervous system 119
    cell death in 130
    induction of 43
    reeler mutant 129
neural connections, ordering of 127
neural crest
    differentiation of 97–98
    migration 25, 27
neural tube 18, 127
neuronal development
    and critical period 132
    and experience 133
neurons 119
    and adhesiveness 127
    birthday of 127–8
    and connections to muscle 123
    and factors 130
    growth of 121
    migration of 121
    migration in cortex 128

network of 129
    and pathfinding 123
neurulation 18–19
Newton, I. 133
Nobel Prize 43, 132, 173
nuclear determinants 33
nuclear transplantation 85–7
nucleic acids 8, 80
nucleotides 79–80
nucleus 7, 77, 80
    interaction with cytoplasm 84
    reactivation of 84
Nüsslein-Volhard, C. 110

oligodendrocyte 100
ommatidium 114
oncogenes 163
ontogeny, and phylogeny 185
optic nerve, regrowth of 126
optic tectum 124
organization, levels of 9
origami 17, 32
organizer 43–4
    and hydra 175

pair-rule genes 107, 109
panda's 'thumb' 154
pangenes 136
paternal genes, requirement for 141
pathfinding, by neurons 121
pattern formation 31
patterning, small scale of 72
phylogeny, and ontogeny 185
phylotypic stage 185–6
pigment cells 27
pigment pattern 57
placenta 141, 145
polarity
    hydra 174
    fruit fly 107
    mouse embryo 36
    sea-urchin 35
polarizing region 61, 63–4
    discovery of 73
positional fields, size of 39
positional information 37, 39
    and fly egg 109
    in leg and antenna 113
    and limb 60
    and regeneration 175, 177–9
    and timing 65
positional signals 43, 47
    and limb 61
positional value, and growth 154
preformation 2, 32–3
prepattern 68

programme
  descriptive 199
  developmental 17
  generative 17, 199
  of growth 148
  internal of cell 47
progress zone 61, 65–6
promoter 83, 90
protein 8, 79–80
  housekeeping 79
  luxury 79, 91
protein synthesis, control of 83
protozoa, and regeneration 181
pygmies 152

quail marker 98

radiation sickness 95
raction-diffusion 55–6, 69, 107
red blood cell, maturation of 92
reeler 129
regeneration 171
  not adaptive 176
  and ciliates 182
  and complete circle rule 179
  and limb 177–9
  and positional information 175, 177–80
  and regulation 181
  and retinoic acid 179
regulation 34, 40
  mouse embryo 36
  and regeneration 181
relative growth 153
repeated units 21
restriction point, and cell cycle 158
retina 124
  and connection to brain 125
retinoic acid
  and limb 64–5
  and regeneration 179
RNA 80
Roux, W. 33

Saunders, J. 72
sea squirt 53
sea-urchin 13, 39
segmentation, fruit-fly 107, 109
  vertebrate 21
selection, and instruction 47
self-assembly 24
self-organization 34, 54–5, 69
sevenless 115
sex 135
  and brain development 142
  genetic determination of 137–8

secondary characters 139
  and hormones 137–8
  temperature control of 141
sexual behaviour 142
sexual characters 137–8
sexual development
  abnormal 141
  environmental determination of 141
sheets, folding 18–21
shell
  coiling of 50
  and evolution 190
sickle cell anaemia 81
signals
  positional 43, 63
  repeated use of 201
  simplicity of 48
  stop and go 47
skin 97
skull, growth of baboon's 154
slime mould 27
'small is beautiful' 200
somites 21
Spemann, H. 43–4
sperm 136
squint 132
Sperry, R. 124
spina bifida 18
spiral cleavage, and snails 50
sponge, and self assembly 24
stem cells 94–101
Swammerdam, J. 3

teeth 18, 20
  and induction 46
tendons 70–1
testicular feminization 140
testis 137–9
testosterone 137–8
  and bird song 142
test-tube, and fertilization 137
thalidomide 67
Thompson, D'A. 190
time-lapse film 13
timing 58
  and evolution 192
tinkering, and evolution 187
transgenic mouse 89
  and nervous system 129
transcription 83–4
Trembley, A. 173
trophoblast 48–9
tunicate 53
  and cytoplasmic localization 54
tumours (see cancer)
Turing, A. 54, 69

turtles and sex 141
twins, human 36

universal principles 10

Valéry, P. 202
variety of development 49
  amphibian 45
vegetal pole, sea-urchin 35
vertebrae 26
vertebrates, similarity of embryos 183
virgin birth 141–2
vis vitalis 5
Vitalist 34
von Baer, K. 183

Waddington, C. 91
Weismann, A. 32–3, 135
Werner's syndrome 169
white cells 93
wing 60

X chromosome 137–8
  inactivation of 101
  and origin tumours 160
X-irradiation 95, 97

Y chromosome 137–8
yolk 11, 145